PRAISE FOR

The Male Brain

"This book is full of surprises, and the reader comes away with more respect for nature in all its intricate wisdom and intelligence."
—*New York Times Style Magazine*

"With her frank talk about sex, delivered unblinkingly and with a smile, coupled with her encyclopedic exposition on the brain, from its medial preoptic area to the temporal parietal unction, Brizendine comes across as part modern-day Dr. Ruth, part nerdy scientist."
—*San Francisco Chronicle*

"In this utterly fascinating follow-up to her bestselling *The Female Brain*, Harvard neuropsychiatrist Brizendine leads readers through the lifespan of a man's brain, using lively prose and personable anecdotes to turn complex scientific research into a highly accessible romp."
—*Publishers Weekly*, starred review

“Brizendine personalizes every developmental stage of male neuroendocrinology and shows how knowing what’s necessarily male about one’s husband, son, father, and other male intimates facilitates better living and working with them . . . Excellent on its topic, not least because it seems as authoritative as it is quick reading.” —*Booklist*

“Dr. Brizendine has marshaled a host of impressive data and insights and presented them in an elegant and entertaining way to clearly illustrate men’s reality—as infants, boys, teens, lovers, husbands, fathers, and workers. It’s a deep dive into the worlds of men, as well as a fascinating read. And along the way, you will pick up some valuable tips to help you understand, appreciate, and connect with the men in your life.”

—Helen Fisher, Ph.D., author of *Why Him? Why Her?*

“The remarkable brain science behind Mars and Venus is a really enjoyable read! I think that this book, along with *The Female Brain,* should be read by every parent, child, husband, wife, employer, employee, and dating-age adult—they bring love and understanding into our most important, and sometimes most frustrating, relationships.”

—Martin L. Rossman, M.D., clinical faculty, UCSF; founder of TheHealingMind.org; clinical professor of medicine at the University of California, San Francisco

The Male Brain

Also by Louann Brizendine, M.D.

The Female Brain

The Male Brain

Louann Brizendine, M.D.

THREE RIVERS PRESS
New York

This book is not intended to take the place of medical advice from a trained medical professional. Readers are advised to consult a physician or other qualified health professional regarding treatment of their medical problems. Neither the publisher nor the author takes any responsibility for any possible consequences from any treatment, action, or application of medicine, herb, or preparation to any person reading or following the information in this book.

Published in the United States by Three Rivers Press,
an imprint of the Crown Publishing Group,
a division of Random House, Inc., New York.
www.crownpublishing.com

THREE RIVERS PRESS and the Tugboat design are
registered trademarks of Random House, Inc.

Originally published in hardcover by Broadway Books, an imprint of the Crown Publishing Group, a division of Random House, Inc., New York, in 2010.

Library of Congress Cataloging-in-Publication Data
Brizendine, Louann, 1952–
The male brain / Louann Brizendine.
p. cm.
1. Neuroendocrinology. 2. Brain—Sex differences.
3. Men—Physiology. I. Title.
QP356.4.B75 2010
612.8—dc22 2009042116

ISBN 978-0-7679-2754-3
eISBN 978-0-307-58939-2

PRINTED IN THE UNITED STATES OF AMERICA

Illustrations by Martie Holmer
Cover photograph/illustration by Viktor Koen
Cover design by Jean Traina

3 5 7 9 10 8 6 4

First Paperback Edition

To the men in my life:

My husband, Dr. Samuel Herbert Barondes

My son, John "Whitney" Brizendine

My brother, William "Buzz" Brizendine II

And in memory of my father, Reverend William Leslie Brizendine

CONTENTS

ACKNOWLEDGMENTS

This book had its beginnings during my educational years at U.C. Berkeley, Yale, Harvard, and U.C. London, so I would like to thank those teachers who most influenced my thinking during those years: Frank Beach, Mina Bissell, Harold Bloom, Marion Diamond, Walter Freeman, Florence Haseltine, Richard Lowenstein, Daniel Mazia, Fred Naftolin, Stanley Jackson, Roy Porter, Carl Salzman, Leon Shapiro, Rick Shelton, Gunter Stent, Frank Thomas, George Valliant, Clyde Willson, Fred Wilt, Richard Wollheim.

During my years on the faculty at Harvard and UCSF, my thinking has been influenced by: Cori Bargman, Samuel Barondes, Sue Carter, Regina Casper, Lee Cohen, Mary Dallman, Allison Doupe, Deborah Grady, Mel Grumbach, Leston Havens, Joel Kramer, Fernand Labrie, Sindy Mellon, Michael Merzenich, Joseph Morales, Kim Norman, Barbara Parry, Victor Reus, Eugene Roberts, Nirao Shah, Carla Shatz, Stephen Stahl, Marc Tessier-Lavigne, Rebecca Turner, Owen Wolkowitz, Chuck Yingling, and Ken Zack.

My colleagues, staff, residents, medical students, and patients in the Women's and Hormone Clinic. I would especially like to thank my faculty members at the clinic: Lyn Gracie Adams, Steve Hamilton, Dannah Hirsch, Jane Hong, Shana Levy, Faina Novosolov, and Elizabeth Springer.

And for their friendship and support throughout: Lynne Benioff, Marc Benioff, Diane Cirincione, Janet Durant, Adrienne Larkin, Sharon Melodia, Nancy Milliken, Jeanne Roberston, Sandy Robertson, Alla Spivak, and Jodi Yeary.

The work presented in this book has been based on and greatly benefited from the research and writings of: Marty Altemus, Arthur Arnold, Arthur Aron, Simon Baron-Cohen, Andreas Bartels, Frank Beach, Jill Becker, Sherri Berenbaum, Karen Berkley, Jeff Blaustein, Marc Breedlove, Lucy Brown, David Buss, Larry Cahill, Anne Campbell, Sue Carter, David Crews, Susan Davis, Karl Deisseroth, Geert De Vries, Catherine Dulac, Elisa Epel, Helen Fisher, David Geary, Jay Giedd, Jill Goldstein, Louis Gooren, Mel Grumbach, Andy Guay, Elizabeth Hampson, Bob Handa, James Herman, Melissa Hines, Gert Holstege, Sarah Hrdy, Janet Hyde, Tom Insel, Bob Jaffe, Doreen Kimura, Eleanor Maccoby, Dev Manoli, Helen Mayberg, Martha McClintock, Erin McClure, Bruce McEwen, Michael Meaney, Toni Pak, Barbara Parry, Don Pfaff, David Rubinow, Robert Sapolsky, Peter Schmidt, Nirao Shah, Barbara Sherwin, Elizabeth Spelke, Dick Swaab, Jane Taylor, Shelley Taylor, Rebecca Turner, Kristin Uvnas-Moberg, Victor Viau, Myrna Weissman, Sandra Witelson, Sam Yen, Kimberly Yonkers, Elizabeth Young, Larry Young, and the many other scientists whose work I have cited in this book.

I would also like to thank the foundations and supporters of my work: the Lynne and Marc Benioff Family, the Lawrence Ellison Medical Foundation, National Center for Excellence in Women's Health at UCSF, the Osher Foundation, the Staglin Family Music Festival for Mental Health, the Salesforce.com Foundation, the Stanley Foundation, and the UCSF Department of Psychiatry.

This book was written and rewritten with the assistance of Toni Robino. I owe her the greatest debt of gratitude.

I would especially like to thank Diane Middlebrook and the Literary Salon. Diane set the stage for me to begin writing; she read many drafts of my work and was, and is, an inspiration even past her untimely death.

Amy Hertz believed in this book from day one and deserves special thanks for helping shape my thinking and writing over the years.

I am very thankful to those who worked to make this book happen: Julie Sills, Stephanie Bowen, Elizabeth Rendfleisch, Mark Birkey, Gary Stimeling, Lorraine Glennon, Diane Salvatore, my ever encouraging agent, Lisa Queen of Queen Literary, and my dedicated publicity manager at Random House, Rachel Rokicki.

I am grateful to my editor at Random House, Kris Puopolo, who supported me with intelligence, skill, and dedication through many years of writing, rewriting, starts, and stops.

I also want to thank my son, John "Whitney," for graciously allowing me to use many of his personal stories and whose help in understanding the world of boys, teens, and young men has been invaluable. His sense of humor and determination continue to inspire me.

Most of all I thank my husband and soul mate, Sam Barondes, for everything—his insights into the world of men, his wisdom, levity, intelligence, critiques, editorial advice, scientific acumen, tolerance, empathy, and love.

The Male Brain

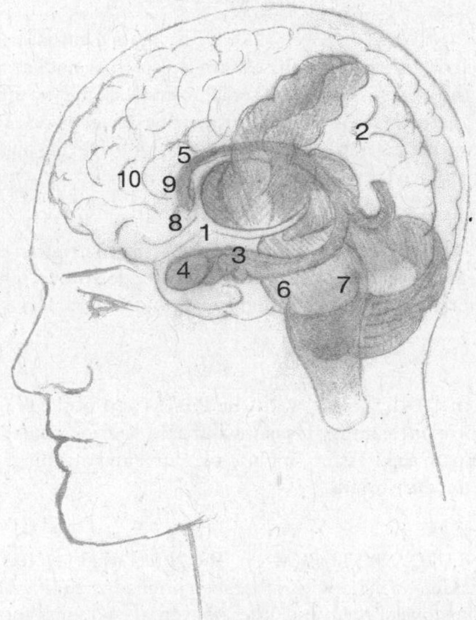

Scientists think of brain areas like the ACC, TPJ, and RCZ as being "hubs" of brain activation, sending electrical signals to other areas of the brain, causing behaviors to occur or not occur.

1. MEDIAL PREOPTIC AREA (MPOA): This is the area for sexual pursuit, found in the hypothalamus, and it is 2.5 times larger in the male. Men need it to start an erection.

2. TEMPORAL PARIETAL JUNCTION (TPJ): The solution seeker, this "cognitive empathy" brain hub rallies the brain's resources to solve distressing problems while taking into account the perspective of the other person or people involved. During interpersonal emotional exchanges, it's more active in the male brain, comes on-line more quickly, and races toward a "fix-it-fast" solution.

3. DORSAL PREMAMMILLARY NUCLEUS (DPN): The defend-your-turf area, it lies deep inside the hypothalamus and contains the circuitry for a male's instinctive one-upmanship, territorial defense, fear, and aggression. It's larger in males than in females and contains special circuits to detect territorial challenges by other males, making men more sensitive to potential turf threats.

4. AMYGDALA: The alarm system for threats, fear, and danger. Drives emotional impulses. It gets fired up to fight by testosterone, vasopressin, and cortisol and is calmed by oxytocin. This area is larger in men than in women.

5. ROSTRAL CINGULATE ZONE (RCZ): The brain's barometer for registering social approval or disapproval. This "I am accepted or not" area keeps humans from making the most fundamental social mistake: being too different from others. The RCZ is the brain center for processing social errors. It alerts us when we're not hitting the mark in our relationship or job. During puberty, it may help males reset their facial responses to hide their emotions.

6. VENTRAL TEGMENTAL AREA (VTA): It's the motivation center—an area deep in the center of the brain that manufactures dopamine, a neurotransmitter required for initiating movement, motivation, and reward. It is more active in the male brain.

7. PERIAQUEDUCTAL GRAY (PAG): The PAG is part of the brain's pain circuit, helping to control involuntary pleasure and pain. During sexual intercourse, it is the center for pain suppression, intense pleasure, and moaning. It is more active during sex in the male brain.

8. MIRROR-NEURON SYSTEM (MNS): The "I feel what you feel" emotional empathy system. Gets in sync with others' emotions by reading facial expressions and interpreting tone of voice and other nonverbal emotional cues. It is larger and more active in the female brain.

9. ANTERIOR CINGULATE CORTEX (ACC): It's the worry-wart, fear-of-punishment area and center of sexual performance anxiety. It's smaller in men than in women. It weighs options, detects conflicts, motivates decisions. Testosterone decreases worries about punishment. The ACC is also the area for self-consciousness.

10. PREFRONTAL CORTEX (PFC): The CEO of the brain, the PFC focuses on the matter at hand and makes good judgments. This "pay total attention to this now" area also works as an inhibiting system to put the brakes on impulses. It's larger in women and matures faster in females than in males by one to two years.

THE CAST OF NEUROHORMONE CHARACTERS

(how hormones affect a man's brain)

TESTOSTERONE—Zeus. King of the male hormones, he is dominant, aggressive, and all-powerful. Focused and goal-oriented, he feverishly builds all that is male, including the compulsion to outrank other males in the pecking order. He drives the masculine sweat glands to produce the come-hither smell of manhood—androstenedione. He activates the sex and aggression circuits, and he's single-minded in his dogged pursuit of his desired mate. Prized for his confidence and bravery, he can be a convincing seducer, but when he's irritable, he can be the grouchiest of bears.

VASOPRESSIN—The White Knight. Vasopressin is the hormone of gallantry and monogamy, aggressively protecting and defending turf, mate, and children. Along with testosterone, he runs the male brain circuits and enhances masculinity.

MÜLLERIAN INHIBITING SUBSTANCE (MIS)—Hercules. He's strong, tough, and fearless. Also known as the Defeminizer, he ruthlessly strips away all that is feminine from the male. MIS builds brain circuits for exploratory behavior, suppresses brain circuits for female-type behaviors, destroys the female reproductive organs, and helps build the male reproductive organs and brain circuits.

OXYTOCIN—The Lion Tamer. With just a few cuddles and strokes, this "down, boy" hormone settles and calms even the fiercest of beasts. He increases empathic ability and builds trust circuits, romantic-love circuits, and attachment circuits in the brain. He reduces stress hormones, lowers men's blood pressure, and plays a major role in fathers' bonding with their infants. He promotes feelings of safety and security and is to blame for a man's "postcoital narcolepsy."

PROLACTIN—Mr. Mom. He causes sympathetic pregnancy (couvade syndrome) in fathers-to-be and increases dads' ability to hear their babies cry. He stimulates connections in the male brain for paternal behavior and decreases sex drive.

CORTISOL—The Gladiator. When threatened, he is angry, fired up, and willing to fight for life and limb.

ANDROSTENEDIONE—Romeo. The charming seducer of women. When released by the skin as a pheromone he does more for a man's sex appeal than any aftershave or cologne.

DOPAMINE—The Energizer. The intoxicating life of the party, he's all about feeling good, having fun, and going for the gusto. Excited and highly motivated, he's pumped up to win and driven to hit the jackpot again and again. But watch out—he is addictively rewarding, particularly in the rough-and-tumble play of boyhood and the sexual play of manhood, where dopamine increases ecstasy during orgasm.

ESTROGEN—The Queen. Although she doesn't have the same power over a man as Zeus, she may be the true force behind the throne, running most of the male brain circuits. She has the ability to increase his desire to cuddle and relate by stimulating his oxytocin.

PHASES OF A MALE'S LIFE

Hormones can determine what the brain is interested in doing. Their purpose is to help guide social, sexual, mating, parenting, protective, and aggressive behaviors. They can affect being rough-and-tumble, competing in sports or attending sporting events, solving problems, interpreting facial expressions and others' emotions, male-male bonding, dating and mating, ogling attractive females, forming sexual and pair-bond relationships, protecting family and turf, fantasizing, masturbating, and pursuing sex.

	MAJOR HORMONE CHANGES	WHAT MALES HAVE THAT FEMALES DON'T
FETAL GROWTH	Brain development: starting 8 weeks after conception, testosterone *masculinizes* and then works with MIS hormone to *defeminize* the male brain.	Y chromosome.
BOYHOOD	Continued production of MIS; low levels of testosterone during this "juvenile pause."	High testosterone from 1 month to 12 months after birth; lower testosterone from 1 to 11 years old; continued high MIS hormone; low estrogen.
PUBERTY	20-fold increase in testosterone along with increasing vasopressin; low MIS.	Increased sensitivity and growth of sexual-pursuit circuits and territorial aggression.
SEXUAL MATURITY, SINGLE MAN	Testosterone continues to be high and activate mating, sex, protection, hierarchy, and turf circuits.	Focused on curvaceous, fertile females. Wants sex first, then love and relationship *may* follow; high libido.
FATHERHOOD	During the mother's pregnancy and after birth of baby, prolactin goes up, testosterone goes down.	Male-pregnancy or couvade syndrome.
MIDLIFE MANHOOD	Very gradually decreasing testosterone.	Continued focus on sex, turf, and attractive women.
ANDROPAUSE	Gradually lower testosterone; by age 85, testosterone level is less than half of what it was at age 20.	Can continue to reproduce; continued focus on sex and attractive women.

MALE-SPECIFIC BRAIN CHANGES	REALITY CHANGES
Growth and masculinization of circuits for sexual pursuit, exploratory behavior, and rough-and-tumble muscle movements.	
More brain circuits for exploratory behavior, rough-and-tumble muscle movements; circuits for male sexual activity continue to develop.	Major interest in winning, movement, chasing objects, rough-and-tumble and exploratory play with boys, not girls.
Circuits for visual sex attraction focus on female figures; perceives male faces as hostile; sense of smell for pheromones changes; auditory perception changes; circuits for sleep cycle change.	Major interest in turf, social interaction, girls' body parts, sexual fantasy, masturbation, male hierarchy; goes to sleep and gets up later; avoidance of parents, challenges authority.
Visual circuits change to spot fertile females and potentially aggressive males.	Major interest in finding sexual partners; focus on job, money, and career development.
Circuits for sex drive suppressed due to lower testosterone and high prolactin; auditory circuits enhanced for hearing babies cry; father-baby synchrony develops.	Major focus is on protecting the mother and baby, on making a living and supporting the family; hears babies cry better than non-dads.
Slowly decreasing activation by testosterone and vasopressin.	Major focus is on raising kids, power and status at work; less focus on must-have-sex-now.
Brain circuits usually fueled by testosterone and vasopressin are declining; ratio of estrogen to testosterone increases; higher oxytocin.	Major interest is in staying healthy and improving well-being, marriage, sex life, grandchildren, legacy; closest that men will ever come to being like women, since oxytocin makes them more open to affection and sentiment, and declining testosterone makes them less aggressive.

The Male Brain

INTRODUCTION

What Makes a Man

YOU COULD say that my whole career prepared me to write my first book, *The Female Brain.* As a medical student I had been shocked to discover that major scientific research frequently excluded women because it was believed that their menstrual cycles would ruin the data. That meant that large areas of science and medicine used the male as the "default" model for understanding human biology and behavior, and only in the past few years has that really begun to change. My early discovery of this basic inequity led me to base my career at Harvard and the University of California–San Francisco (UCSF) around understanding how hormones affect the female and male brains differently and to found the Women's Mood and Hormone Clinic. Ultimately that work led me to write *The Female Brain*, which addressed the brain structures and hormonal biology that create a uniquely female reality at every stage of life.

The distinct brain structures and hormonal biology in the

male similarly produce a uniquely male reality. But as I considered writing *The Male Brain*, nearly everyone I consulted made the same joke: "That will be a short book! Maybe more of a pamphlet." I realized that the idea that the male is the default-model human still deeply pervades our culture. The male is considered simple; the female, complex.

Yet my clinical work and the research in many fields, from neuroscience to evolutionary biology, show a different picture. Simplifying the entire male brain to *just* the "brain below the belt" is a good setup for jokes, but it hardly represents the totality of a man's brain. There are also the seek-and-pursue baby boy brain, the must-move-or-I-will-die toddler brain; the sleep-deprived, deeply bored, risk-taking teen brain; the passionately bonded mating brain; the besotted daddy brain; the obsessed-with-hierarchy aggressive brain; and the fix-it-fast emotional brain. In reality, the male brain is a lean, mean problem-solving machine.

The vast new body of brain science together with the work I've done with my male patients has convinced me that through every phase of life, the unique brain structures and hormones of boys and men create a male reality that is fundamentally different from the female one and all too frequently oversimplified and misunderstood.

Male and female brains are different from the moment of conception. It seems obvious to say that all the cells in a man's brain and body are *male*. Yet this means that there are deep differences, at the level of every cell, between the male and female brain. A male cell has a Y chromosome and the female does not. That small but significant difference begins to play out early in the brain as genes set the stage for later amplification by hormones. By eight weeks after conception, the tiny male testicles begin to produce enough testosterone to marinate the brain and fundamentally alter its structure.

Over the course of a man's life, the brain will be formed and re-formed according to a blueprint drafted both by genes and male sex hormones. And this male brain biology produces his distinctly male behaviors.

The Male Brain draws on my twenty-five years of clinical experience as a neuropsychiatrist. It presents research findings from the advances over the past decade in our understanding of developmental neuroendocrinology, genetics, and molecular neuroscience. It offers samplings from neuropsychology, cognitive neuroscience, child development, brain imaging, and psychoneuroendocrinology. It explores primatology, animal studies, and observation of infants, children, and teens, seeking insights into how particular behaviors are programmed into the male brain by a combination of nature and nurture.

During this time, advances in genetics, electrophysiology, and noninvasive brain-mapping technology have ignited a revolution in neuroscientific research and theory. Powerful new scientific tools, such as genetic and chemical tracers, positron-emission tomography (PET) and functional magnetic resonance imaging (fMRI), now allow us to see inside the working human brain while it's solving problems, producing words, retrieving memories, making decisions, noticing facial expression, falling in love, listening to babies cry, and feeling anger, sadness, or fear. As a result, scientists have recorded a catalog of genetic, structural, chemical, hormonal, and brain-processing differences between women and men.

In the female brain, the hormones estrogen, progesterone,

and oxytocin predispose brain circuits toward female-typical behaviors. In the male brain, it's testosterone, vasopressin, and a hormone called MIS (Müllerian inhibiting substance) that have the earliest and most enduring effects. The behavioral influences of male and female hormones on the brain are major. We have learned that men use different brain circuits to process spatial information and solve emotional problems. Their brain circuits and nervous system are wired to their muscles differently—especially in the face. The female and male brains hear, see, intuit, and gauge what others are feeling in their own special ways. Overall, the brain circuits in male and female brains are very similar, but men and women can arrive at and accomplish the same goals and tasks using different circuits.

We also know that men have two and a half times the brain space devoted to sexual drive in their hypothalamus. Sexual thoughts flicker in the background of a man's visual cortex all day and night, making him always at the ready for seizing sexual opportunity. Women don't always realize that the penis has a mind of its own—for neurological reasons. And mating is as important to men as it is to women. Once a man's love and lust circuits are in sync, he falls just as head over heels in love as a woman—perhaps even more so. When a baby is on the way, the male brain changes in specific and dramatic ways to form the daddy brain.

Men also have larger brain centers for muscular action and aggression. His brain circuits for mate protection and territorial defense are hormonally primed for action starting at puberty. Pecking order and hierarchy matter more deeply to men than most women realize. Men also have larger processors in the core of the most primitive area of the brain, which registers fear and triggers protective aggression—the amygdala. This is why some men will fight to the death defending their loved ones. What's more, when faced with a loved one's emotional distress, his brain area for problem solving and fixing the situation will immediately spark.

I must have been dimly aware of this long catalog of distinctive male behaviors when I first found out, twenty-one years ago, that the baby I was carrying had a Y chromosome. I immediately thought, *Oh dear. What am I going to do with a boy?* Up until that moment, I realized, I had unconsciously been thinking *It's a girl!* and feeling confident that my own female life experiences could guide me in raising a daughter. I was right to be nervous. My lack of boy-smarts was about to matter more than I imagined. I now know from my twenty-five years of research and clinical work that both men and women have a deep misunderstanding of the biological and social instincts that drive the other sex. As women, we may love men, live with men, and bear sons, but we have yet to understand men and boys. They are more than their gender and sexuality, and yet it is intrinsic to who they are. And it further complicates matters that neither women nor men have a good sense of what the others' brains or bodies are doing from one moment to the next. We are mostly oblivious to the underlying work performed by different genes, neurochemicals, and hormones.

Our understanding of essential gender differences is crucial because biology does not tell the whole story. While the distinction between boy and girl brains begins biologically, recent research shows that this is *only* the beginning. The brain's architecture is not set in stone at birth or by the end of childhood, as was once believed, but continues to change throughout life. Rather than being immutable, our brains are much more plastic and changeable than scientists believed a decade ago. The human brain is also the most talented learning machine we know. So our culture and how we are taught to behave play a big role in shaping and reshaping our brains. If a boy is raised to "be a man," then by the time he becomes an adult, his brain's architecture and circuitry, already predisposed that way, are further contoured for "manhood."

Once he reaches manhood, he will likely find himself pondering an age-old question: What do women want? While no one has a definitive answer to that question, men do know what women and society in general want and expect from *them.* Men must be strong, brave, and independent. They grow up with the pressure to suppress their fear and pain, to hide their softer emotions, to stand confidently in the face of challenge. New research shows that their brain circuits will architecturally change to reflect this emotional suppression. Although they crave closeness and cuddling as much or perhaps even more than women, if they show these desires, they are misjudged as soft or weak by other men and by women, too.

We humans are first and foremost social creatures, with brains that quickly learn to perform in socially acceptable ways. By adulthood, most men and women have learned to behave in a gender-appropriate manner. But how much of this gendered behavior is innate and how much is learned? Are the miscommunications between men and women biologically based? This book aims to answer these questions. And the answers may surprise you. If men and women, parents and teachers, start out with a deeper understanding of the male brain, how it forms, how it is shaped in boyhood, and the way it comes to see reality during and after the teen years, we can create more realistic expectations for boys and men. Gaining a deeper understanding of biological gender differences can also help to dispel the simplified and negative stereotypes of masculinity that both women and men have come to accept.

This book provides a behind-the-scenes brain's-eye view of little boys, tumultuous teens, men on the mating hunt, fathers, and grandfathers. As I take readers through the phases of the male brain's life, my hope is that men will gain a greater understanding of their deepest drives and women will catch a glimpse of the world through male-colored glasses. We are entering an era, finally, when both men and women can begin to understand their distinct biology and how it affects their lives. If we know how a biological brain state is guiding our impulses, we can choose how to act, or to not act, rather than merely following our compulsions. If you're a man, this knowledge can not only help you understand and harness your unique male brain power, but it can also help you understand your sons, your father, and other men in your life. If you're a woman, this book will help you interpret and comprehend the intricacies of the male brain. With that new information, you can help your sons and husbands to be truer to their nature and feel more compassionate toward your father.

Over the years, as I have been writing this book, I have come to see the men I love most—my son, my husband, my brother, and my father—in a new light. It is my hope that this book will help the male brain to be seen and understood as the fine-tuned and complex instrument that it actually is.

ONE

The Boy Brain

DAVID RACED past the swing set and zoomed around the toolshed in the backyard with his preschool buddies Matt and Craig hot on his heels. Determined to maintain his lead, he took a shortcut through the sandbox, sending sand and shovels flying as he made a beeline for the coveted Big Wheel tricycle. Matt pushed Craig aside and dived for the wheeled wonder, but David was already sliding into the driver's seat. With pedals churning, David screeched off down the sidewalk and into the driveway, where he victoriously spun doughnut after doughnut.

Disappointed but not to be outdone, Matt and Craig headed for the open garage to see what else they could find to ride. Craig spotted it first: a large plastic trash can. "Let's use this!" he shouted. And without another word, the boys were running headlong for the hill in the backyard, dragging the can behind them. "C'mon. Gimme a push!" Craig commanded as he slid into the can. "Harder!" he said, as Matt's first shove barely budged it. Matt rammed the can with his shoulder as hard as he could, and the green vehicle tumbled down the hill with Craig inside whooping and hollering.

You don't have to study brain science to know that little boys are all about action and adventure. Go to a playground and you'll see boys like David and his friends in perpetual motion. Boys are programmed to move, make things move, and watch things move. Scientists used to think this stereotypical boy behavior was the result of socialization, but we now know that the greater motivation for movement is biologically wired into the male brain.

If you watched the fetal development of a male and a female brain with a miniature time-lapse brain scanner, you'd see these critical movement circuits being laid down from the blueprint of their genes and sex hormones. Scientists agree that when cells in various areas of the male and female brains are stimulated by hormones like testosterone and estrogen, they turn on and off different genes. For a boy, the genes that turn on will trigger the urge to track and chase moving objects, hit targets, test his own strength, and play at fighting off enemies.

David and his friends weren't taught to be action-oriented; they were following their biological impulses. David's mother said that his love affair with movement was obvious from day one. "When I put him in his bassinet, I thought he'd cry and look beseechingly at me the way Grace did when she was a baby," she said. "But as soon as he spotted the moving mobile, he forgot I was there."

David was only twenty-four hours old, and without encouragement or instruction from anyone, he stared at the rotating triangles and squares on the mobile and seemed to find them fascinating. Nobody taught David to follow the movements of the dangling triangles and squares with his eyes. He just did it. A boy's superior ability to track moving objects isn't the result of being conditioned by his environment. It's the result of having a male brain. Every brain is either male or female and, while they are mostly alike, scientists have discovered some profound differences. Certain behaviors and skills are wired and programmed innately in boys' brains, while others are wired innately in girls'. Scientists have even found that male-specific neurons may be directly linked to stereotypical male behaviors like roughhousing. And studies show that from an early age, boys are interested in different activities than girls. These differences are reinforced by culture and upbringing, but they begin in the brain.

WHAT MAKES A BOY A BOY?

I met David's mother, Jessica, a few months after he was born. Her daughter, Grace, was three years old, and Jessica and her husband, Paul, were thrilled to have a beautiful baby boy. But Jessica was worried because things weren't going quite as smoothly with David as they had with Grace. Jessica said, "He's sweet and cuddly one minute, and the next minute he's squirming out of my arms. If I don't put him down, he shrieks like I'm killing him."

Jessica was afraid that David might be hyperactive. But her pediatrician told her David was just fine and developing normally. Researchers at Harvard found that baby boys get emotionally worked up faster than girls, and once they're upset, they're harder to soothe. So, early on, parents spend more time trying to dial down their sons' emotions than their daughters'.

She said, "Grace was easier to calm. David keeps us constantly on our toes!"

Jessica also told me that David didn't make eye contact with her the way Grace did when she was a baby. She said that he'd only look at her for a couple of seconds and then go right back to staring at the mobile. I couldn't help but smile, because I had this same concern with my own son. At that time, psychologists believed the key to developing a bond with your baby was what they called mutual gazing—looking into each other's eyes. Whereas that's true for baby girls, it turns out that baby boys bond without as much mutual gazing. And unlike girls, who are inclined to look long and hard at faces, boys' visual circuits pay more attention to movement, geometric shapes, and the edges and angles of objects from the get-go.

I said to Jessica, "By the time they're six months old, baby girls are looking at faces longer and making eye contact with just about everyone. But baby boys are looking *away* from faces and *breaking* eye contact much more than girls. There's nothing wrong with David. His brain just doesn't find eyes and faces as interesting as toy airplanes and other moving objects."

David's male brain was prompting him to visually explore animated objects. We now know that genes on the Y chromosome are the reason. Like other boys, David's fascination with movement was the result of circuitry that started to form in his brain just eight weeks after he was conceived. During fetal development, David's brain was built in two stages. First, during weeks eight to eighteen, testosterone from his tiny testicles *masculinized* his body and brain, forming the brain circuits that control male behaviors. As his brain was marinating in testosterone, this hormone began to make some of his brain circuits grow and to make others wither and die.

Next, during the remaining months of pregnancy another hormone, MIS, or Müllerian inhibiting substance, joined with testosterone and *defeminized* David's brain and body. They suppressed his brain circuits for female-type behaviors and killed off the female reproductive organs. His male reproductive organs, the penis and testicles, grew larger. Then, together with testosterone, MIS may have helped form David's larger male brain circuits for exploratory behavior, muscular and motor control, spatial skills, and rough play. Scientists discovered that when they bred male mice to lack the MIS hormone, they did not develop male-typical exploratory behavior. Instead, they behaved and played more like females. The female brain circuits that make a girl a girl are laid down and develop without the effects of testosterone or MIS.

After I shared this information with Jessica, she raised her eyebrows and asked, "Are you saying that if Grace's brain had been exposed to these male hormones when I was pregnant, she'd act more like David?"

"That's right," I said, smiling as her face lit up with recognition. It's always rewarding to me when I see this kind of relief on a mother's face. Suddenly, instead of thinking that she's doing something wrong or that there's something wrong with her child, she can relax and begin to appreciate her son's maleness.

She said, "It's just so different with David. He's so much more active than Grace was, even at this age. But he can be the very essence of sweetness, too.

"The other day when I was having a hard time getting him down for his nap, Paul took him and played with him on our bed, hoping he'd calm down. I had my doubts about whether it would work, but when I peeked in to check on them a little later, David was lying with his tiny hand inside of Paul's big one, and they were both fast asleep."

From birth until a boy is a year old, a period that scientists call infantile-puberty, his brain is being marinated in the same high levels of testosterone as in an adult man. And it's this testosterone that helps stimulate a boy's muscles to grow larger and improves his motor skills, preparing him for rough-and-tumble play. After the year of infantile-puberty, a boy's testosterone drops, but his MIS hormone remains high. Scientists call this period, from age one to ten, the juvenile pause. They believe that the MIS hormone may form and fuel his male-specific brain circuits during this ten-year period, increasing his exploratory behavior and rough play. This meant it wouldn't be long before Jessica would have more reason to worry as David started testing his limits, as I well remember with my own son.

When he was a toddler and we were out walking on Baker Beach in San Francisco, he took off running after a sandpiper toward the water. I shouted and waved my arms like a madwoman to signal danger. He completely ignored me. I had to run after him and grab his shoulders to pull him back from the surf, just as a huge wave was rolling in. That was the first day in what would be years of his ignoring my signals of danger—stop, don't do that—requiring me to keep a firm hold on him.

Researchers have found that by the time a boy is seven months old, he can tell by his mother's face when she's angry or afraid. But by the time he's twelve months old, he's built up an immunity to her expressions and can easily ignore them. For girls, the opposite happens. A subtle expression of fear on Jessica's face would stop Grace in her tracks. But not David.

By the age of one, David seemed oblivious to the look of warning on Jessica's face. Researchers asked mothers of one-year-old boys and girls to participate in an experiment in which an interesting but forbidden toy was placed on a small table in the room with them. Each mother was told to signal fear and danger with only her facial expressions, indicating that her child should not touch it. Most of the girls heeded their mother's facial warning, but the boys seemed not to care, acting like they were magnetically pulled toward the forbidden object. Their young male brains may have been more driven than the girls' by the thrill and reward of grabbing the desired object, even at the risk of punishment. And this also happens with fathers. In another study, with dads and their one-year-olds, the boys tried to reach forbidden objects more often than the girls. The fathers had to give twice as many verbal warnings to their sons as to their daughters. And researchers found that by the age of twenty-seven months, boys more often than girls will go behind their parents' backs to take risks and break rules. By this age, the urge to pursue and grab items that are off-limits can become a hair-raising game of hide-and-seek—with parents hiding the danger their sons will inevitably seek.

When David was three and a half, Jessica told me that he never ceased to amaze her, both for better and for worse. "He picks me flowers, tells me he loves me, and showers me with kisses and hugs. But when he gets the urge to do something, the rules we've taught him vanish from his mind." She told me that David and his friend Craig were in the bathroom washing up for dinner when she heard Craig yell, "Stop it, David. I'm peeing." Then she heard the distinct sound of the hair dryer. *Danger* flashed through Jessica's brain. Racing down the hall, she flung open the bathroom door just in time to get a splash of urine on her legs. David had turned the blow-dryer on his friend's stream—just to see what would happen. But being sprayed with urine didn't upset her nearly as much as David's disregarding the "no electrical appliances without adult supervision" rule. For the next couple years, she had to keep all electrical appliances well out of David's reach. But, she told me with a slight blush, "There's one thing I can't keep out of his reach, even in public."

PLAYING WITH HIS PENIS

David thought nothing of grabbing and playing with his penis—anytime, anywhere. A boy's public relationship with his penis is something that has made many mothers wince, including me. But the male brain's reward center gets such a huge surge of pleasure from penis stimulation that it's nearly impossible for boys to resist, no matter what their parents threaten. So rather than trying to stop David, I suggested Jessica start teaching him to explore this compelling pleasure in the privacy of his room.

A few weeks after Jessica started trying to get David to play with his penis in "privacy," the family went on vacation. As they were walking down the hallway in their hotel, David saw a sign hanging on the doorknob of the room next-door and asked, "Mom, what does P-R-I-V-A-C-Y say?" When Jessica said the word out loud for him, he said, "Oh, that man must be doing *his privacy* in there." From then on, he'd refer to playing with his penis as "doing my privacy."

BOYS' TOYS

Later that year, when David came into the office with Jessica, I handed him a lavender toy car from an assortment I had in a shoebox. He frowned as he said, "That's a girl car." Tossing the car back into the box, he grabbed the bright red car with black racing stripes, saying, "This is a boy one!" Researchers have found that boys and girls both prefer the toys of their own sex, but girls will play with boys' toys, while boys—by the age of four—reject girl toys and even toys that are "girl colors" like pink.

I didn't know this when my own son was born, so I gave him lots of unisex toys. When he was three and a half years old, in addition to buying him one of the action combat figures he was begging for, I bought him a Barbie doll. I thought it would be good for him to have some practice playing out nonaggressive, cooperative scenarios. I was delighted by how eagerly he ripped open the box. Once he freed her from the packaging, he grabbed her around the torso and thrust her long legs into midair like a sword, shouting, "Eeeehhhg, take that!" toward some imaginary enemy. I was a little taken aback, as I was part of the generation of second-wave feminists who had decided that we were going to raise emotionally sensitive boys who weren't aggressive or obsessed with weapons and competition. Giving our children toys for both genders was part of our new child-rearing plan. We prided ourselves on how our future daughters-in-law would thank us for the emotionally sensitive men we raised. Until we had our own sons, this sounded perfectly plausible.

Scientists have since learned that no matter how much we adults try to influence our children, girls will play house and dress up their dollies, and boys will race around fighting imaginary foes, building and destroying, and seeking new thrills. Regardless of how we think children should play, boys are more interested in competitive games, and girls are more interested in cooperative games. This innate brain wiring is apparently different enough that behavioral studies show that boys spend 65 percent of their free time in competitive games, while girls spend only 35 percent. And when girls are playing, they take turns twenty times more often than boys.

It is commonly said that "boys will be boys," and it's true. My son didn't turn Barbie into a sword because his environment promoted the use of weapons. He was practicing the instincts of his male brain to aggressively protect and defend. Those stereotypically girl toys I gave him in his first few years of life did not make his brain more feminine any more than giving boy toys to a girl would make her more masculine.

I later found out that my son had plenty of masculine company when it came to turning Barbie into a weapon. In an Irish nursery school, researchers observed that boys raided the girls' kitchen toys and even unscrewed the faucet handles in the miniature sink to use as toy guns. In another nursery-school study, researchers found that preschool boys were six times more likely than girls to use domestic objects as equipment or weapons. They used a spoon as a flashlight to explore a make-believe cave, turned spatulas into swords to battle the "bad guys," and used beans as bullets.

The next time I talked to Jessica, she told me David came home from kindergarten one day with a black eye. His teacher said he had called Craig a sissy for playing with the girls, and Craig hauled off and hit him. Jessica said, "I felt so bad for him that I took him out for ice cream, and out of the blue, he turned to me and said, 'I love you, Mommy. I'm gonna marry you when I grow up.' Seeing him with that black eye and hearing him say that to me just about broke my heart. Why would his best friend hit him like that, just for calling him a name?"

I told Jessica that by the time a boy is just three and a half, the greatest insult is being called a girl. Boys tease and reject other boys who like girls' games and toys. And after the age of four, if a boy plays with girls, the other boys soon reject him. Studies show that beginning in the toddler years, boys develop a shared understanding about which toys, games, and activities are "not male" and must therefore be avoided. Boys applaud their male playmates for male-typical behavior while they condemn everything else as "girly."

Curiosity about the origins of boys' strong preference for masculine toys led researchers to explore this further with young rhesus monkeys. Because monkeys are not gender socialized as to which toys are masculine or feminine, they made good subjects for this study. Researchers gave the young male and female monkeys a choice between a wheeled vehicle, the "masculine" toy, and a plush doll figure, the "feminine" toy. The males almost exclusively spent time playing with the wheeled toy. But the females played equal amounts of time with the doll and the wheeled toy. The scientists concluded that gender-specific toy preferences have roots in the male brain circuitry in both boys and male monkeys. And there is further evidence that this toy preference has its origins in fetal brain development. In human girls, a prenatal exposure to high testosterone, due to a disorder called CAH (congenital adrenal hyperplasia), has been found to influence later toy preferences. By the time these CAH girls are three or four, they prefer boy-typical toys more than other girls do.

Scientists believe that boys' toys reflect their preference for using big muscle groups when they play. A related preference for action shows up even in art class. Researchers found that elementary-school boys preferred to draw action scenes like car and plane crashes. Nearly all their drawings captured a dynamic movement, and they used only a few colors. The girls in the study drew people, pets, flowers, and trees and used many more colors than the boys did.

David not only liked drawing action scenes and playing with boy toys, but by the age of five, his favorite board game was Chutes and Ladders. He would do anything to win, including cheat. He'd slyly move his marker the wrong number of spaces so he could climb up a ladder or avoid having to slide down a chute. And he was devastated when he lost. Jessica said, "Every time Craig and David play this game, they end up fighting." I could relate. When my son was in kindergarten, we had to remove all the win-lose board games and put them in the closet for a while. Victory is critically important to boys because, for them, play's real purpose is to determine social ranking. At an early age, the male brain is raring for play-fighting, defending turf, and competing. Losing is unacceptable. To a young male brain, the victory cry is everything.

PUSHING HIS LIMITS

"Aarghhh!" David shouted as he charged forward, thrusting and jabbing his new laser sword at Craig. Not to be outdone, Craig snatched the sword out of David's hands and took off running with it. But he made it only a few yards before David caught up and grabbed the back of his mud-caked shirt. Within seconds they were on the ground wrestling for possession of the sword. To someone not familiar with the ways of young boys, this would look like a fight. But David and Craig were having a blast.

Boys wrestle and pummel each other with gusto, competing for toys and trying to overpower each other. They play this way up to six times more than girls do, a reality that Jessica now found highly entertaining, although she hadn't always seen the humor in it. Boys discover their place in the world by pushing all of their body's physical limits, so it's not just fighting but also being able to fart or burp the loudest or the longest that gives a boy bragging rights. Jessica said, "I'll never understand why David and Craig think farting on each other is so funny. But they think it's hilarious, and Paul laughs as hard as *they* do."

For David and Craig, every day was filled with a series of serious physical contests. How fast can you run? How high can you climb? How far can you jump? A boy's success or failure in sports and other contests can make or break his sense of self. Even though Jessica could appreciate that males are naturally driven to test their physical abilities, she still worried that David would get hurt. But Paul—who grew up with three brothers—knew that the bumps and bruises were a normal part of boyhood.

During the juvenile pause, boys imitate their dads, uncles, and older male cousins, and they're particularly intrigued with the men who stand out as alpha males. Go to the zoo and watch the primates, and you'll see the most powerful male sitting by himself chewing grass and the little guys running up and attacking him from behind. The little guys are playing at things they'll be required to do in their future. When the alpha male has had enough, he'll shoo away the juveniles.

Undaunted, they will continue to wrestle with each other, literally tumbling across the ground. This rough-and-tumble play is also observed in groups of human boys everywhere.

SHOWING HIS STRENGTH

By the time boys are in first grade, they get a brain high when they show their strength and aggression. Using physical force together with insults is even better. As child researcher Eleanor Maccoby says, "These boys are just trying to have their kind of fun." This way of playing gives their brain a massive feel-good reward in the form of a dopamine rush. The neurochemical dopamine is addictively rewarding—the brain likes it and wants more—so boys are always seeking the thrill of the next high. That's why they love scary movies, haunted houses, and daring each other to take risks. Boys don't necessarily want to get hurt, but they usually think the excitement is worth it. Jessica said, "I'm just happy to get through a day without putting ice or Band-Aids on somebody."

By grade school, the play styles of boys and girls in groups have diverged, and children self-impose sex segregation. Observational studies found that, worldwide, boys on playgrounds wrestle, roughhouse, and mock-fight frequently; girls do not. In addition to their different play styles, boys and girls may also dislike playing together because, as research shows, by the time boys are in first grade, they're no longer paying much attention to girls or listening to what they say. A study of boys in a first-grade classroom in Oregon found that boys paid the most attention first and foremost to what other boys said. The teachers placed second, and the girls placed a distant third—if they placed at all. As a matter of fact, ignoring girls altogether was the most common. David and most of the other boys in his first-grade class had already sworn off playing with girls, and their female classmates were just fine with that. They didn't like playing with the boys either.

A study on an Irish kindergarten playground may shed even more light on the girls' and boys' interactions with each other. The researchers noted that the boys monopolized the tricycles and bicycles and played ramming games, while girls—on the few occasions they got a turn to ride—were very careful not to hit other kids' bikes or anything else. The boys even became territorial and possessive of their bikes, showing a willingness to fight for them that the girls did not show.

FIRST IN LINE

Jessica said she couldn't understand it when David's teacher wrote on his report card that he was always fighting to be the first in line for recess and lunch. Since Grace never seemed to mind waiting her turn in line, the importance David placed on being first took Jessica by surprise.

The pecking order clearly matters more to boys. Studies show that by age two, a boy's brain is driving him to establish physical and social dominance. And by the age of six, boys tell researchers that *real fighting* is the "most important thing to be good at." Scientists have also learned that boys are remarkably fast at establishing dominance in a group through rough-and-tumble play.

In a study conducted with boys and girls at a nursery school, the boys demonstrated a clear hierarchy by the end of their first play session. Among the girls, some dominance hierarchy was established too, but it was more fluid. In the boy groups, however, by the end of just the second play session, the boys unanimously agreed about the ranking position of each boy, and these rankings remained stable for the remainder of the six-month study.

How do boys know so quickly who's tough and who's not? While bigger boys typically rank higher in status, researchers found that the leaders weren't always the biggest. In the study, the alpha boys were the ones who refused to back down during a conflict. These boys aggressively demonstrated their strength by picking on, intimidating, or roughing up boys who challenged them. In the hormone tests taken on all the boys in the group, it turned out that alpha boys had higher testosterone levels than did the other boys. And to the researchers' surprise, the rank a boy had attained in the group by the age of six predicted where he'd be in the hierarchy at age fifteen.

Of course, only one boy can be the top dog, so the rest must find other ways to succeed and avoid being picked on in the boy pack. One strategy is to form an alliance with the alpha boy by giving him things he wants and doing him favors. When my son was in elementary school, he casually asked me to buy him the biggest bags of Chex Mix to send to school with him for snack time. I thought he wanted to share them with his friends, so I didn't question it. It wasn't until I inadvertently bought him the smaller size that I discovered why he'd wanted the big bag. It turned out that he'd been using Chex Mix at recess "to hire everyone he could hire," as he put it or, as I saw it, to buy off the top dogs and appease the bullies. When he saw the smaller bag on the counter by his backpack, he shouted, "Now I'm done for! And all because of *you*!"

Boys can usually work things out within the checks and balances of the boy pack, but this cruel *Lord of the Flies* system still strikes horror in most mothers' hearts—including mine. Regardless of how mothers feel about it, though, boys instinctively know they must learn how to succeed within the male hierarchy. And that's not the only type of learning boys do differently.

SQUIRMING BOYS LEARN BETTER

Tightly clutching the remotes in their fists, David and Craig punched, jabbed, and dodged, occasionally throwing an insult along with a punch. As with many boys their age, Wii had become their favorite toy. To use this active video-game system, the boys mimicked the action they wanted to see displayed on the screen. When David threw a punch, his video character mirrored him. When Craig dodged the punch, his character did the same.

Research from Stanford University showed that playing Wii activates parts of the male brain linked to dopamine production. Boys get rewarded by this feel-good brain chemical, just as they do when they're roughhousing. The more opponents they conquer, the more stimulated their male brain becomes, and the more dopamine their brains release. It's a thrill a minute.

Even in a conventional video game, when a boy is not actually moving, watching every move of the athlete or video character still gives him a thrill. Moreover, the signal gets sent from his brain through the neurons and into the muscles in his body even if he isn't moving. If we were to watch David's body and brain with an fMRI camera when he plays a game like Super Mario Brothers, every time he makes Mario jump, we'd see David's brain activate the neurons that control his own jumping muscles. He would embody the movement he sees even though he's not really jumping. Boys react more physically to their environment than girls do in this way. Their muscles are practically twitching in response to everything they see going on around them. And that difference may mean that boys use their muscles and nervous systems more than girls to think and express themselves as well.

For instance, when a boy first learns to read the word *run*, his brain fires messages to his leg muscles and makes them twitch: He's rehearsing the action of running in order to learn the word. And to read and understand the meaning of the word *slug*, David's sensation area in the brain for slimy and squishy is activated. Then the movement area of his brain for slow and slithering is engaged, and even the emotional area of his brain for disgust gets into the action. These brain areas are needed for him to completely embody, learn, and remember the meaning of *slug*. Scientists refer to this process as embodied cognition, because the muscles and body parts he uses to learn a word will stay connected to the meaning of that word. This is true for all our brains but seems particularly significant for boys. It may annoy their teachers, but boys who squirm can learn better than boys who sit still.

Boys like David are twisting and turning all the time, and scientists believe this may also give them their advantage at spatial manipulation. By age five, according to researchers in Germany, boys are using different brain areas than girls to visually rotate an object in their mind's eyes. The boys mentally rotated the pictures of the objects by using both sides of their brain's spatial-movement area in the parietal lobe. Girls used only one side to do the task. While that in itself is revealing, what I found most intriguing is that this spatial-movement area in the male brain is locked in the "on" position. That means it's always working in the background on autopilot. But in the female brain, this parietal area is "off," waiting in standby mode, and not turned on until it's needed.

From age five on, mental rotation of objects is one of the biggest cognitive differences between boys and girls. In the boy brain, solving problems that require spatial rotation begins in the visual cortex and goes straight to the already "on" parietal spatial-movement area in both hemispheres. It then fires signals to the muscles that cause them to mimic the shape and position of the object. The researchers concluded that most boys, and also some girls, get a holistic, visceral sense of how an object occupies space—they embody its reality, making it easier for them to grasp its three-dimensionality.

Curious to find out how this applies practically in the classroom setting, researchers studied students in a grade-school math class to see how girls and boys solved conceptual math problems and how long it took them. The boys solved the problems faster than the girls. But what was most surprising to the researchers was that most of the boys, when asked to explain how they got the answer, gave an explanation without using any words. Instead, they squirmed, twisted, turned, and gestured with their hands and arms to explain how they got the answer. The boys' body movements *were* their explanations. Words, in this instance, werc a hindrance.

What also got my attention about this study was what the researchers did next with the girls. In the following six weeks of the experiment, they taught the girls to explain their answers with the same muscle movements the boys had made without using words. At the end of the six weeks, once the girls stopped talking and started twisting and turning, they solved the problems as quickly as the boys. The male and female brains have access to the same circuits but, without intervention, use them differently.

THAT BOY SMELL

Around age eleven, the juvenile-pause stage of a boy's life begins winding down. One of the most pungent signs that he is entering the next stage is the new scent he starts to exude. It's not BO yet; it's more like sweaty socks. When my son was this age, we mothers referred to it as "that boy smell": not quite the musk of manhood but no longer the sweet smell of childhood. What we smelled was the male sweat glands under the influence of testosterone giving off small amounts of the pheromone called androstenedione. This increase in testosterone signaled the dawn of puberty.

A rise in testosterone sparks a new interest in girls—or at least in their female body parts. That curiosity is what got David into trouble at school in fifth grade. David's fourteen-year-old cousin texted him a photo of a bare-breasted woman, and the boys in his class were huddling around to sneak a peek. Earlier that day, they'd all been disappointed in their sex-ed class's lack of detailed information. This was more like it. Even though sex-hormone levels are quite low during most of the juvenile pause, once boys approach puberty, they begin to relentlessly pursue every scrap of sexual information they can get their hands on.

When David's teacher called, Jessica was upset, but when she told Paul what happened, he felt a little surge of "that's my boy" pride and couldn't help cracking a smile. While Jessica thought this was a big deal, Paul knew that looking at naked photos was tame compared with what David would soon be doing. When his male hormones turned back on and the juvenile pause ended, Paul and Jessica would have more to worry about than David's sexual curiosity. Soon, his action, exploration, and risk-taking brain circuits would be running at high speed, urging him to prove himself again and again. The anger and aggression circuits that were formed before he was born and strengthened during boyhood were about to be hormonally fuel-injected.

When that happens, every trait and tendency set up in his male brain during childhood—action, strength, desire for dominance, exploring, and taking risks—will be magnified. His brain circuitry and rising hormone levels will cause him to question and disobey his parents, seek sexual partners, strike out on his own, fight for his place in the male hierarchy, find a mate, and come into his own by entering manhood. With testosterone driving his reality, he will soon feel strong, brave, and invincible. Feeling cocksure, he will be blind to consequences and deaf to his parents' warnings.

TWO

The Teen Boy Brain

"Turn off your computer *now*, Jake! No gaming until that homework is done!" shrieked Jake's mother as she pounded on his bedroom door. Opening the door a crack, Jake gave her a blank stare and grumbled something under his breath before shutting the door in her face. Kate knew he'd probably turn the computer back on without the volume. But what she didn't know was that free porn sites were beginning to be more enticing to him than the war games he played online with his buddies.

Kate was a patient of mine, and up until this past year, she'd described her relationship with Jake as close and rewarding. But when her formerly happy and cooperative son turned fourteen, he became sullen and irritable. Struggle, struggle, struggle is all they seemed to do these days. When Kate and her husband, Dan, found out that Jake hadn't turned in a single English assignment in weeks, they worried that he might be drinking or experimenting with drugs. That's when they scheduled a family appointment with me. During our session, Jake stared out the window and Dan listened politely as Kate wrenchingly complained that their son had suddenly become unreachable and secretive. Not only had Jake gotten into a fight with another freshman, named Dylan, but he also had a new group of friends, including a girl named Zoe whom Kate described as "fast." Dan spoke up in disagreement, saying, "I'm not too worried about the fight or Jake's new friends. But I *do* expect Jake to keep his grades up."

Meanwhile, Jake, with his curly brown hair and long, lanky limbs, seemed dazed and oblivious to his parents' worries about him. When I turned and asked him, "What do you think of your parents' concerns?" he merely shrugged. It was clear that Jake, like most teens, wasn't going to say much of anything in front of his parents, so I suggested that he come in for a private session the following week. Since my own teen son had recently left for college after four long years of high school, I had a pretty good idea what Jake and his parents were going through. No matter how harmonious a boy's childhood has been, puberty can change everything. This stage of child development requires that delicate parental maneuver of becoming disengaged without disengaging. Kate said she felt as if the Jake she knew had disappeared, and in some ways he had.

Scientists have discovered that the teen brain in both sexes is distinctly different from the preadolescent brain. The changes that were becoming obvious in Jake were set in motion by his genes and hormones while he was still in utero. Now, with the end of the juvenile pause, it was time for Jake to ramp up his skills for surviving in a man's world. And he was ready and eager, even if his mother wasn't. At this stage, the millions of little androgen switches, or receptors, in his brain are hungrily awaiting the arrival of testosterone—king of the male hormones. As the floodgates are flung wide open, the juice of manhood saturates his body and his brain. When my own son turned fourteen and became moody and irritable, I remember thinking, "Oh my God, soon the testosterone will take him over mind, body, and soul."

TESTOSTERONE TSUNAMI

Although Kate worried that Jake's behavior was extreme, I assured her that he was no different from many other boys his age. At fourteen, Jake's brain would have already been under reconstruction for a few years. Between the ages of nine and fifteen, his male brain circuitry, with its billions of neurons and trillions of connections, was "going live" as his testosterone level soared twentyfold. If testosterone were beer, a nine-year-old boy would get the equivalent of about one cup a day. But by age fifteen, it would be equal to two *gallons* a day. Jake wasn't into drugs or alcohol. He was loaded on testosterone.

From then on, testosterone would biologically masculinize all the thoughts and behaviors that emerge from his brain. It would stimulate the rapid growth of male brain circuits that were formed before he was born. It also would enlarge his testicles, activate the growth of his muscles and bones, make his beard and pubic hair grow, deepen his voice, and lengthen and thicken his penis. But just as dramatically, it would make his brain's sexual-pursuit circuits, in his hypothalamus, grow more than twice as large as those in girls' brains. The male brain is now structured to push sexual pursuit to the forefront of his mind.

Early in puberty, when images of breasts and other fe-

Testosterone in a Male's Life

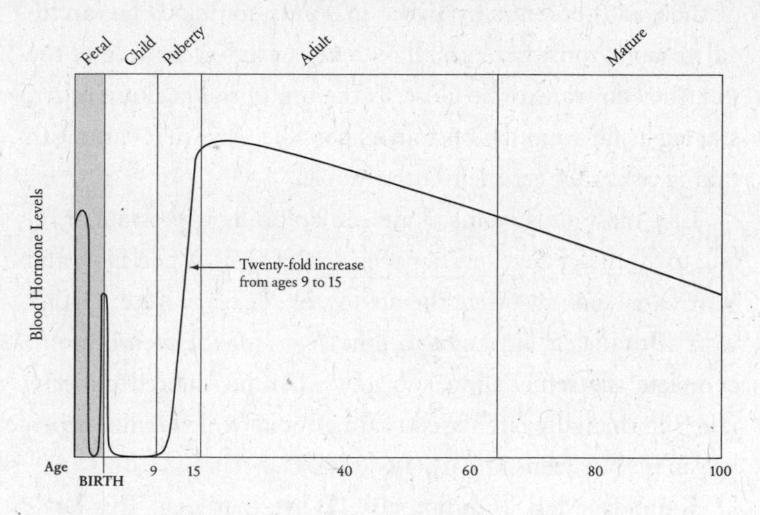

male body parts naturally take over their brain's visual cortex, some boys wonder if they're turning into "pervs." It takes a little while for them to get used to their new preoccupation with girls, which runs on autopilot. This sexual preoccupation is like a large-screen TV in a sports bar—always on in the background. When I share this information with teen boys in high-school classrooms, I can see recognition flash across many of their faces, if only for an instant, before they go back to looking bored.

But sex is not the only thing on a teen boy's mind. As the testosterone surged through Jake's brain cells, it was stimulating a companion hormone called vasopressin. Together, testosterone and vasopressin were making Jake's brain territorial about his room and sensitive to his peer's putdowns—perceived or real. And when these hormones got mixed with the stress hormone cortisol, they supercharged his body and brain, preparing him for the male fight-or-flight response in reaction to challenges to his status or turf. Our brains have been shaped for hundreds of thousands of years by living in status-conscious hierarchical groups. And while not all teen boys want to be king of the hill, they do want to be close to the top of the pecking order, staying as far from the bottom as possible. And that can mean taking risks that get them into trouble.

Like most of us moms, Kate couldn't fully appreciate or relate to all the changes in her teen son's brain. When Dan and Kate came into my office the next week, I said to Kate, "Don't worry. It takes about eight to nine years for the teen brain to complete the remodeling it began when he entered puberty. Jake's hormonally enhanced brain circuits will stabilize when he's in his late teens or early twenties."

Kate's face fell. "I'm not sure I'll live that long. This boy's killing me." I could see that she was only half joking.

Dan turned to me and said, "Look, Jake's just like every other teenage boy that ever walked the planet Earth. He's gonna look at some porn. He's gonna blow off his homework, get in some fights, and drool over girls. Once he's grounded for a while, he'll come around."

HOMEWORK WARS

Even though Jake was now grounded until he completed every neglected English assignment, it was still hard for him to focus his brain on schoolwork. If we could watch Jake's brain with a miniature brain scanner as he sat down to do homework, we'd see his prefrontal cortex, or PFC—the area for attention and good judgment—flickering with activity as it tried to force him to focus on his studies. We'd also see bursts of vasopressin and testosterone pulsing through his brain, activating his sex and aggression circuits. When an image of Dylan's smirking face registered in Jake's brain, his stress hormone, cortisol, would start climbing. Now his threat and fear center—the amygdala—would activate. And then, as an image of Zoe in that tight sweater she'd been wearing in class today flashed across his secondary visual system, we'd see his sexual circuits activate, distracting him further. Next we'd see his PFC struggling to regain focus on his English homework. But it would be too late. His PFC was no match for his sexual daydreams. Soon, homework would be the last thing on his mind.

Teen boys aren't trying to be difficult. It's just that their brains aren't yet wired to give much thought to the future. Getting boys to study and do homework has always been more of a battle for parents than getting girls to do the same, and with today's high-tech temptations, the battle can feel like a war. Studying instead of doing something fun online just doesn't make sense to teen boys. Research shows that it takes extraordinarily intense sensations to activate the reward centers of the teen boy brain, and homework just doesn't do it. Fortunately for Jake, his father came up with both a stick and a carrot—the threat of being grounded without his computer, cell phone, or TV for the next month versus a pair of tickets to a playoff game if he maintained a B average and turned in all his homework. I have to admit I was a little surprised when Jake's grades immediately improved. Somehow Jake took his dad's threat and reward to heart.

I knew that even boys who are getting good grades can begin to hate school by the time they're in tenth or eleventh grade. So the next time I met with Jake, I asked him if there was anything at all that he looked forward to at school. He raised his eyebrows as if I must be joking and said, "No. We're not allowed to leave the building or even open our cell phones on campus. It's so stupid. It's like jail." I could see that this was going to be a tough year for Jake and for his parents. Everything about our school system is in direct conflict with teen boys' adventurous, freedom-seeking brains. So we shouldn't be surprised that boys cause 90 percent of the disruptions in the classroom or that 80 percent of high-school dropouts are boys. Boys get 70 percent of the Ds and Fs. They're smart enough to get good grades, but soon they just don't care. And it doesn't help that school start times are totally out of sync with the teen brain's sleep cycle.

SLEEPY AND BORED

Jake had English class first thing in the morning and said it took everything he had just to stay awake. He said, "I never fall asleep before two in the morning. On weekends I sleep late, but it pisses off my mom."

The sleep clock in a boy's brain begins changing when he's eleven or twelve years old. Testosterone receptors reset his brain's clock cells—in the suprachiasmatic nucleus, or SCN—so that he stays up later at night and sleeps later in the morning. By the time a boy is fourteen, his new sleep set point is pushed an hour later than that of girls his age. This chronobiological shift is just the beginning of being out of sync with the opposite sex. From now until his female peers go through menopause, he'll go to sleep and wake up later than they do.

Nowadays, most teen boys report getting only five or six

hours of sleep on school nights, while their brains require at least ten. Some parents have to unplug the Internet if they want their sons to get any sleep at all. If school systems and teachers really wanted teenagers to learn, they'd make start times later by several hours. At least that would increase the chances of a boy's eyes being open—even if it wouldn't wipe the look of boredom from his face.

Like many parents, I used to think teen boys were *acting* bored because it was no longer cool to be excited about anything. But scientists have discovered that the pleasure center in the teen boy brain is nearly numb compared with this area in adults and children. The reward center in Jake's brain had become less easily activated and wasn't sensitive enough to feel normal levels of stimulation. He wasn't acting bored. He *was* bored, and he couldn't help it. When Erin McClure and colleagues at the National Institute of Mental Health scanned teenagers' brains while they looked at shocking pictures of grotesque and mutilated bodies, their brains didn't activate as much as children's or adults'. As many high-school teachers know, the teen boy brain needs to be more intensely scared or shocked to become activated even the tiniest bit. The amount of stimulation it takes to make an adult cringe will barely get a rise out of a teen boy. If you want to startle them enough to make them scream or jump, you'll have to magnify the experience with sounds, lights, action, and gore. Now I know why my son liked the bloodiest special effects and shoot-'em-up movies when he was a teen. This preference may not change as boys reach manhood, as blockbuster moviemakers well know. But grown men don't need the same raw rush as they did when they were thrill-seeking teens.

Jake's mom blamed his glazed-over eyes, irritability, and

short fuse on lack of sleep, and that definitely had something to do with it. But what she didn't know was that a lot of Jake's anger was being triggered by the new way his male brain was experiencing the world and everyone in it.

SEEING THE WORLD THROUGH MALE-COLORED GLASSES

If a woman could see the world through "male-colored glasses," she'd be astonished by how different her outlook would be. When a boy enters puberty and his body and voice change, his facial expressions also change, and so does the way he perceives other people's facial expressions. Blame it on his hormones. A key purpose of a hormone is to prime new behaviors by modifying our brain's perceptions. It's testosterone and vasopressin that alter a teen boy's sense of reality. In a similar fashion, estrogen and oxytocin change the way teen girls perceive reality. The girls' hormonally driven changes in perception prime their brains for emotional connections and relationships, while the boys' hormones prime them for aggressive and territorial behaviors. As he reaches manhood, these behaviors will aid him in defending and aggressively protecting his loved ones. But first, he will need to learn how to control these innate impulses.

Over the past year, for no good reason, Jake began to feel much more irritable and angry. He would quickly jump to the conclusion that people he encountered were being hostile toward him. We might ask, Why did it seem the whole world suddenly turned on him? Unbeknownst to Jake, vasopressin was hormonally driving his brain to see the neutral faces of others as unfriendly. Researchers in Maine tested teens' perceptions of neutral faces by giving them a squirt of vasopressin nasal spray. They found that, under the influence of this hormone, the teen girls rated neutral faces as more friendly, but the boys rated the neutral faces as more unfriendly or even hostile. This may explain why the next time Jake saw Dylan, he thought his face looked angry when, in fact, Dylan was just bored. Jake's brain was being primed by hormones for getting into trouble.

In animals that are in puberty, scientists have discovered that priming the males' brains with vasopressin and testosterone changes their behavior, too. The scientists found that the brain's two master sensors for emotions—the amygdala and hypothalamus—became supersensitive to potential threats when hormonally primed. And in animal studies in which male voles were given vasopressin, it resulted in more territorial aggression and mate protection.

In humans, a potential threat is often signaled by a facial expression. Before puberty, when Jake had less testosterone and vasopressin, Dylan's bored face probably wouldn't have looked hostile or angry to him. But now everything was different. Evolutionary biologists believe seeing faces as angrier than they actually are serves an adaptive purpose for males. It allows them to quickly assess whether to fight or to run.

At the same time, Jake and Dylan were also honing the ancient male survival skills of facial posturing and bluffing. They were learning to hide their emotions. Some scientists believe human males have retained beards and facial hair, even in warmer climates, in order to make them look fierce and hide their true emotions.

In the male hierarchy of primates and humans, the angry face is used to maintain power. And the angriest faces typically belong to men with the highest testosterone, according to research. A study of teen boys in Sweden found that the ones with the most testosterone reacted more aggressively to threats. These boys with the highest testosterone also reported being more irritable and impatient. And in another study, testosterone levels rose in response to seeing an angry face, thus dialing up the brain circuits for aggression. So angry faces—real or imagined—ignite the male fighting spirit. As Jake and Dylan had experienced in their shoving match, this sudden anger can trigger a knee-jerk reaction—often surprising even to the fighters. If these two boys had lower testosterone and vasopressin, they would not have been so fired up to fight and wouldn't have felt compelled to even the score. But as it was, this hormone cocktail was keeping an irritable and sometimes irrational fire smoldering.

TUNING OUT

The teen male not only sees faces differently than he did as a boy; he also begins to perceive voices and other sounds differently than he did before adolescence. And his changing hormones can make him hear things differently than girls his age. In Portugal, researchers found that during puberty, estrogen surges in females and testosterone surges in males increase the hearing differences between girls' and boys' brains, but the main difference is that some simple sounds, like white noise, are processed differently in the male brain. Liesbet Ruytjens and colleagues in the Netherlands compared the brain activity of seventeen- to twenty-five-year-old males and females as they processed the sound of white noise and as they processed the sound of music. The female brains intensely activated to both the white noise and to the music. The male brains, too, activated to the music, but they *deactivated* to the white noise. It was as if they didn't even hear it. The screening system in their male brains was automatically turning off white noise. Scientists have learned that during male fetal brain development, testosterone affects the formation of the auditory system and the connections within the brain, making it inhibit unwanted "noise" and repetitious acoustic stimuli more than the female brain does. I tease my husband that his brain's acoustic system seems to automatically shut down when I start repeating myself—it's registering in his brain as white noise.

Likewise, when Zoe and her friends talked endlessly about movies, fashion, and other girls, their combined voices just sounded like humming and buzzing to Jake's ears. For him and the other guys, following the girls' rapid musical banter was practically impossible. The best they could do was nod their heads and pretend to be listening.

Boys can't understand why girls like to talk and text so much or why they need to share every minute detail. Jake and his friends were more likely to send ultrabrief messages about something "important," like the score of a football game or an estimate of the hot substitute teacher's measurements.

Even though older male and female teens in college have been shown to say about the same number of words a day, researchers found that they're interested in talking at different times and about different topics—boys about games and objects and girls about people and relationships. And these differences, too, may be primed by hormones. James Pennebaker at the University of Texas found that as males were undergoing testosterone treatment over a period of one to two years, in their written communications they began using fewer and fewer words about people and more and more words to talk about objects and impersonal topics. When boys are Jake's age, with their high levels of testosterone, they may not talk much about personal topics. And when it comes to talking to adults—especially his parents—a teen boy's motto is "Give nothing away."

LOOKING GOOD AND SAVING FACE

If you peeked in from the back of his classroom, all the guys in Jake's English class would look about the same. You could hardly tell them apart—their clothes a few sizes too big, sloppily hanging off their bodies, their hair purposely left messed up, their faces marked by unshaven facial hair and pimples. Slouched at their desks with expressions of boredom or disdain, they'd look as though they'd just rolled out of bed—and they had. Everything about a teen boy says he couldn't care less about what other people think of him or how he looks. But in reality, just the opposite is true.

Teens are painfully sensitive to the subtle, and sometimes not so subtle, feedback they get from their peers. Even though Jake's face didn't show it, it was clear to me that he had become more and more obsessed with what his classmates thought about him. At his next appointment, he proudly told me that one of Zoe's girlfriends told him Zoe really liked his hair since he'd let it grow long. And he angrily told me that he wasn't going to attend his usual Friday night poker game, because one of the guys had criticized him for taking so long to play his cards. Neither the compliment nor the criticism would have jiggled his brain circuits at all before puberty. Nowadays, every socially relevant comment or look painfully pierced him, or at least his rostral cingulate zone, or RCZ, an area that acts as the brain's barometer for social approval or disapproval. This "I am accepted or not by others" brain area was in the process of a massive recalibration. Now his friends' approval trumped that of his parents. Evolutionary psychologists theorize that brain circuits like the RCZ developed in primitive societies to keep people from making social mistakes that could result in being ostracized by their clans or tribes. Social acceptance could make the difference between life and death. To teenagers, disapproval from peers feels like death. *Fitting in* is everything.

SHOW OF STRENGTH

When Jake felt dissed or challenged, he couldn't rest until he somehow evened the score and regained some respect. Ever since Dylan shoved him at the game, he daydreamed about beating him up. Dylan had a size advantage, so Jake didn't want to pick a real fight with him. But he felt compelled to best him at something. And until he figured out what that was, he'd just have to play it cool. A teen's self-confidence is directly proportional to how he looks in front of his peers. If he can't be on top, the next best thing is pretending not to care. Thus, Jake was now practicing the posturing techniques that men use to get respect. For males, displaying signs of dominance and aggression is an important way to establish and maintain social hierarchy. Even if Jake really didn't feel all that confident, he wanted to look as though he was in charge and not afraid to fight. But as most men know, a show of anger is just as often only a bluff.

Still, with their high testosterone, increased irritability, and this new urge to be dominant, some teen boys do end up physically testing their place in the dominance hierarchy. So it's not unusual for them to have a face-off with an authority figure—even one of their parents, as I found out. My son and I had our toe-to-toe showdown when he was just shy of his sixteenth birthday. I was awakened at two A.M. on a school night by what sounded like rocket blasts, launched by his gaming-computer. It woke me up from a dead sleep, and I was livid. I stomped downstairs in my pajamas, pounded on his bedroom door, and yelled, "Turn off that computer and give me the power cord *now.*" As he opened the door, he puffed up his chest and leaned toward me with his six-foot frame. "There's no way I'm giving it to you," he said. Surprised by how intimidated I felt, I knew I had to stand my ground. In the firmest voice I could muster, I growled, "Either give me that power cord, or you can forget about getting your driving permit next week." He knew I meant business, so he begrudgingly turned over the cord. For the moment, I had won. But as with Jake, his fight for independence was just beginning.

WINNER TAKES ALL

That fall, Jake's mother called me after a few weeks of football practice to report that Jake's attitude at home had improved dramatically. But when the actual season started, Kate reported that he'd become hard to live with. Researchers have found that testosterone levels increase before a competition, so before a game, Jake's neurochemicals—dopamine, testosterone, cortisol, and vasopressin—were cheering him on and making him feel that his team couldn't lose. He was excited and confident. This prefight high happens not only with athletic events but with any competition that the male brain is participating in or even *just watching.* The more testosterone Jake's body made, the more dopamine and vasopressin his brain made, and the more pumped up he felt, especially when his team was winning. Studies show that winning releases more testosterone than losing, even in sports spectators. Winning is a natural high that acts in the brain a lot like drug addiction because it's such a huge rush. But the minute that something goes wrong, the feel-good chemicals bottom out as hopes of victory are dashed.

When Jake's team lost, he was sullen for days. Even seeing Zoe didn't make him feel much better. Kate said she didn't know what was worse, his doom and gloom when they lost or his cockiness when they won. She said, "When they win, he struts around like a rooster, and when they lose, he waves me off like I'm his servant." Lately, Jake had been staying out past his curfew and ignoring his parents' requests to tell them where he was going. He'd learned the covert art of slipping in and out of the house without seeing or talking to anyone. Consequently, Jake thought he'd become the master of deception. So when he decided to ride into the city on the back of a friend's Harley and told his dad he was catching a movie, he thought he'd adequately covered his tracks. What Jake didn't know was that the teen parent network is faster than broadband. When one of the other mothers saw Jake on the back of a motorcycle twenty miles from home, without a helmet, she immediately called Kate. Jake was busted.

SOCIETY'S PURVEYORS OF NEW IDEAS

Kate was more than disappointed in Jake. She was furious and scared. Where had her parenting gone wrong, she wondered, to make him do something so stupid and dangerous? When they came to my office, Dan told me Jake was merely behaving the way he himself had when he was that age—adventurous and devil-may-care, but Kate was taking his latest stunt personally.

"Jake acts like we're idiots!" Kate blurted out. "Like he's the only one who knows anything. When we try to get him to listen, he just rolls his eyes and says, 'This isn't the dark ages anymore. You have no clue what things are like now.' "

I was well aware of how Kate was feeling. My son often accused me of being from the dinosaur age because, according to him, I knew nothing about today's music, hairstyles, clothes, or Internet sites. In every generation, teens need to reject their parents' ideas in favor of their own. By the time a boy is sixteen or seventeen, he will desperately seek autonomy from his parents. Every cell in his brain seems to cry: "Leave me alone and let me live my own life!"

Jake's intense need for separation and independence was primitive and primal. You can see the same independent, risk-taking behavior in other male primates when they reach puberty. Researchers observe that when some adolescent male monkeys leave their birth troop, they strike out on their own with bravado. Scientists believe that adolescent bravery has contributed mightily to the success of the human species and that the curious, incautious, and flexible nature of the teen brain makes teens society's purveyors of new ideas in every generation. Jake's brain was primed for exploration and programmed to break new ground, even if it meant compromising his personal safety—and his mother's sanity.

As I well know, every mother holds her breath and prays that her teenager doesn't do something foolish and end up getting hurt. But according to studies, when teen boys are in a group, their brains experience excitement and emotional euphoria that makes them more willing to do risky things. That's probably why researchers find that when boys are with peers, they have more car wrecks and generally suffer more negative consequences of unsafe, impulsive choices. And although drug and alcohol abuse is reported to increase when teen boys are together, even without those substances, boys take more chances. In a study of teen drivers, the presence of peers more than doubled the number of risks teenage boys took in a video driving game. They concluded that from the teen years through the early twenties, simply being with friends increases risky decision making. Rental car companies, with their age requirement of twenty-five, know what they're doing.

Jake firmly believed he could make his own good decisions and run his life without the interference of adults. He couldn't accept that his brain was not biologically ready to handle independence. Teen boys are certain they have everything under control. But they don't. As I explained to Jake's parents, teens have two distinct systems running their brains.

The activating system—led by the amygdala—develops first. It is impulsive and gets double the stimulation when he's with his peers. It's like a gas pedal. It accelerates. The second system, the inhibiting system—the prefrontal cortex (PFC)—is like a brake. It carefully thinks things through, weighs the risks, and when working smoothly, it stops us from doing things that are dangerous or stupid. Jay Giedd and colleagues at the National Institute of Mental Health found that the inhibiting system doesn't mature in boys until their early twenties. Jake's inhibiting system was still under construction, so his brain was operating with a gas pedal but faulty brakes. Bottom line: parental controls required.

When Jake came in the next time, I asked him if he'd thought about what could have gone wrong on his late-night motorcycle ride. Flashing his most charming but "knowing" smile, he said, "Nothing bad happened. Why can't everybody just chill?"

I could see that Jake's parents had their work cut out for them. And I knew one of my jobs with Kate was going to have to be to help her tolerate the experimenting that would be a necessary prelude to Jake's independent survival. I could vividly remember the gut-wrenching maternal fear I had experienced during similar episodes when my son was a teen. But at the same time, as I had, Kate would need to accept that certain aspects of her son's life would forever be off-limits to her. Already, touching had been off-limits since he was twelve or thirteen. Researchers have shown that teen boys begin to be repulsed, not only by the proximity of their mother's body, but also by her smell. The scientists speculated that this may have evolved as a protection against inbreeding. For years now, whenever Kate tried to straighten Jake's collar or fix his hair, he'd bat her hand away. As Jake's brain set up these new physical barriers with his mother, it also established firm boundaries around his privacy. He certainly wasn't going to share the details of the intimate journey he was the most anxious about embarking on.

HOT AND BOTHERED

All week Jake had been trying to screw up the courage to ask Zoe out. They'd been hanging out in the same group for most of the year but never went out alone. Now he was trying to find out from her friends whether she liked him "like that," or not. Hanging out with her in the group was no longer enough. He felt as if he would burst if he couldn't be alone with her.

Girls don't fully appreciate the bravery it takes for a guy to risk rejection by asking them out. But teen girls soon notice the new power their budding figures have over boys' brains. Boys usually feel the first stirrings of sexual attraction when they're just eleven or twelve years old and begin to have fleeting sexual fantasies. But this is years before they're ready to pair off, and it's the age when boys begin frequent masturbation. Studies show that from puberty until men's midtwenties, they may need to ejaculate one to three times a day. Girls this age reportedly masturbate an average of less than one time per day. Scientists believe that this frequent sexual stimulation is biologically required to keep young men primed, fertile, and ready to have "real sex" at the first opportunity.

Jake's sexual-interest circuits had been flipped on years ago, and his brain's visual cortex had become naturally but indelibly fixated on breasts and buttocks. He obsessively collected every detail about sex that he could find, and when he was with Zoe, he was so mesmerized by her breasts that he often missed what she was saying. He also found himself losing the fight to resist taking a peek at the forbidden online porn sites. He was compelled to learn everything he could so he'd know what to do when the time finally came to "do it." Although Jake didn't consciously know it, his mating brain was now in charge.

During the teen years, a boy's brain circuits undergo major changes. Some brain areas grow wildly, pulsing with constant activity, while other areas are cut back or redirected. It's as if a new operating system is being installed on his computer. Some programs are being upgraded and some are being deleted. The transition can be rocky at times, but once the new system has taken over, he can begin to use the full force of his male brain circuits. And where will he try out these new powers? Wherever there are attractive, desirable girls.

THREE

The Mating Brain: Love and Lust

THE INSTANT Ryan laid eyes on Nicole, she had his undivided attention. He was at a sports bar watching the basketball playoffs with some rugby teammates, but one look at Nicole, and he forgot all about the game. A twenty-eight-year-old Web designer, Ryan had enough dating experience to know that women with good looks don't always have personalities to match. But she triggered his brain's "must have" sequence, and without another thought, he was on his feet and moving toward her. He noticed that her friend was attractive too, but it was Nicole who took his breath away.

With long blond hair, a petite hourglass figure, and a face that could easily belong to a model, Nicole was well aware of the power of her sexual attractiveness. She was now twenty-six, and she'd been my patient since her rocky teenage years. Men had been drooling over her ever since she turned fifteen, grew breasts, and had her braces removed.

As Ryan watched Nicole, he was practically oblivious to everyone but her. His brain's sexual-pursuit area, in his hypothalamus, lit up like a slot machine. Suddenly, all he could think about was how to get her attention. Without being consciously aware of it, Ryan was following the commands of his ancient mating brain.

The men alive today have been biologically selected over millions of years to focus on fertile females. What they don't know is that they've evolved to zoom in on certain features that indicate reproductive health. Researchers have found that the attraction to an hourglass figure—large breasts, small waist, flat stomach, and full hips—is ingrained in men across all cultures. This shape tells his brain that she's young, healthy, and probably not pregnant with another man's child. Like all men's, Ryan's number-one mate-detection circuit was visual. A male's visual cortex comes prewired to notice women who are shaped like Nicole. Men don't really have one-track minds, but when their brains enter "mate-pursuit mode," they can seem to.

FLIRTING IS A "CONTACT-READINESS" SPORT

When Nicole gave me a detailed recap of their first meeting, it was clear that somehow Ryan had charmed her. If we could have watched the play-by-play of Ryan's nonverbal body movements we would have seen him walk casually but deliberately toward Nicole, hoping she'd look up. Once she did, we'd see him tilt his chin and raise his eyebrows ever so slightly, smiling as he took a step closer. Next, Nicole tipped her head toward him, returned the smile, and leaned back just a little. Her body was saying, *I'm interested, but cautious.* Ryan's mating brain read Nicole right. While still smiling, he took a half-step back.

While Ryan didn't have that chiseled *GQ* look that Nicole found most attractive, he *was* cute and looked harmless enough. His smile and the twinkle in his hazel eyes disarmed her, and she could feel her own smile widening as she looked down to coyly break eye contact.

In scientific lingo, these nonverbal flirting signals that Ryan and Nicole were displaying are called contact-readiness cues. Without saying a word, they were signaling interest to the other's brain. I still smile at the memory of my scientist husband trying to flirt with me and hanging on my every word at the business lunch where we first met. Flirting is a contact-readiness sport, and men who do it best score the most.

As this scene with Ryan and Nicole played out, it might have looked as though their movements and facial expressions had been carefully rehearsed to suit Western culture. But these nonverbal microflirtations appear to come preprogrammed deep in the human brain. Researchers have filmed first encounters between men and women in a variety of cultures and have found that people around the world give the same flirtatious cues as Ryan and Nicole.

While Ryan continued in pursuit mode, he took a deep breath as he quickly screwed up the courage to make his next move, desperately hoping this gorgeous woman was not out of his league. Trying to sound as confident and laid-back as possible, he addressed both Nicole and Maggie: "You two look thirsty. Can I get you some drinks?"

Before Nicole could say no, Maggie accepted the offer. "Thanks! I'd like a glass of Chardonnay. I'm Maggie, by the way, and this is Nicole."

Ryan nodded as he said, "I'm Ryan." Then he turned to Nicole and asked, "And what would you like?"

"Chardonnay sounds good to me too," she said, and Ryan was immediately turned on by the musical quality of her voice.

When he returned with the drinks, Nicole asked, "Are you here to watch the game?"

Ryan, feeling a little more confident now, flashed his most charming smile and said, "No, I'm here to watch you." Nicole was flattered, even though she knew it was just a line. Ryan was playfully incorporating what scientists call little deceits and exaggerations into his flirtation. Researchers found that because men believe women expect flattery from them, they don't see anything wrong with meeting those expectations. While Ryan was comfortable with a few flirtatious exaggerations, he didn't want to push his luck too far, so he casually asked, "Are you a Giants fan or an A's fan?"

"Neither," Nicole said with a grin. "I've been studying for the bar exam, and Maggie threatened to remove me from her friends list if I didn't take a break."

Before Nicole had a chance to say another word, Maggie said, "Pull up a chair, Ryan."

THE MATING SENSES

Ryan had noticed that Maggie's voice was deeper than Nicole's, and while it was pleasing, his brain instantly categorized her as a potential friend, rather than a potential mate. But Nicole's higher-pitched voice had triggered his brain to place her in the "hot and sexy" category.

In a study of an African hunter-gatherer tribe called the

Hadza, men rated women with deeper voices as better foragers, but said they were more sexually attracted to the women with the highest-pitched voices. And the women in the tribe rated the men with the deepest voices as the best hunters and protectors, but were turned off by the men with squeaky or high-pitched voices. Ryan's voice sounded pleasant to Nicole even though it wasn't quite that deep male baritone that made her go weak in the knees.

Now that Ryan was sitting next to Nicole, he was close enough to take in her sweet scent, and his nose instant-messaged his subconscious brain that she not only smelled good, but was also potentially a good genetic match. Our pheromones—odorless "smells" detected by our noses—carry genetic information, according to researchers. Infants who came from the repeated intermarriages of Europe's royal families taught us that couples whose genes are too similar give birth to sickly offspring. And a study in Switzerland of sweaty T-shirts that had absorbed the pheromones of the people who wore them showed that those who were good genetic matches (that is, those who were most dissimilar) smelled best to each other. If Nicole had smelled "bad" to Ryan, he could have been turned off and not even known why. This isn't about hygiene; it's about genes.

Ryan's mating brain was giving him encouraging hormonal signals, and Ryan thought Nicole seemed interested, so he tried to draw her out with another question. "So, when is the bar exam?"

"Next week," she said.

Maggie chimed in, "I'm giving her a bar-exam coming-out party when it's over. Wanna come?"

The party was just a couple of weeks away, but to Ryan

it felt like a month. He couldn't stop thinking about Nicole and found himself silently rehearsing topics he could talk to her about. As it turned out, Ryan didn't have to worry quite so much. The night of the party, the conversation between them was easy, and he felt good that he could frequently make Nicole laugh. And at the end of the party, he was thrilled that she accepted his offer to drive her home. By then, the sexual tension between them was palpable. When he walked her to the door and looked into her eyes, she didn't back away, so he leaned in and kissed her good night. He had intended just a quick kiss, but when their lips met, their tongues followed, seemingly of their own accord. The kiss was so sweet and dizzying that he couldn't break away. Fortunately for Ryan, neither could Nicole.

In the mating game, a kiss is more than a kiss—it's a taste test. Saliva contains molecules from all the glands and organs in the body, so a French kiss serves up our signature flavor. As soon as Ryan's tongue touched Nicole's, information about each other's health and genes was collected and secretly sent to their brains. If Nicole had genes that were too similar to his and the kiss tasted sour, it could have been a sexual deal-breaker. But the kiss was sweet; it led to another and then another. Scientists have learned that there is plenty of bioactive testosterone in men's saliva, enough that it may activate the sexual-arousal center in a woman's brain.

So your mother was right—French kissing *can* lead to sex. Ryan was hoping that tonight was the night, but Nicole gently pulled back, thanked him, and said good night without inviting him in.

SCORING AS SOON AS POSSIBLE

Once Ryan had a taste of Nicole, he was hungry for more. Although he was desperately craving her, he knew he had to wait a few days before calling her, or he'd look too eager. Showing his burning desire wouldn't do him any favors right now, particularly because Nicole was being so cautious. Researchers found that when a man is sexually attracted to a woman, he wants to have sex with her as soon as possible. For the men in the study, waiting a week or more seemed like a very long time. The women, on the other hand, wanted to wait up to three times longer. As much as Ryan was secretly hoping for sooner rather than later, he could tell that Nicole wasn't the type to rush into anything, least of all sex. While that was frustrating on one hand, it was also reassuring and made him think she might be in the long-term mating category.

In the basest way, to a man, winning the mating game means getting his DNA and genes into the next generation. Even though he isn't consciously thinking this, the instinctual part of his brain knows that the more women he has sex with, the more offspring he's likely to have. Meanwhile, the female brain is trying to discern whether a man has what it takes to be a good protector and provider. Researchers find that this holds true regardless of a woman's level of education or financial independence. When Ryan called and invited Nicole out to dinner and a movie, she suggested that they go Dutch—just to eliminate the pressure for sex she sometimes felt when she let men spend money on her. But Ryan wanted to pony up resources to demonstrate how much he valued her and was willing to invest. In studies of mating behavior in primates, biologists have discovered that females have more sex with males who bring them meat. Primatologists have dubbed this the meat-for-sex principle. The males who showed they were willing to provide food got more sexual access to the females, increasing their chances of paternity. Ryan was on the right track.

He liked treating Nicole like a queen, and he had no qualms about spending money on her. But by their fourth date he was burning with sexual desire. He thought he'd better come up with something persuasive fast, or he'd die of sexual frustration. Gifts, flowers, or the promise of a romantic getaway weekend—he was thinking of them all. As men well know, they have to develop and refine an array of smooth tactical maneuvers, because women have different mating goals than men have. The female brain wants the hope of love and commitment before having sex, but for men, sex often comes first. Not surprisingly, it was Ryan's philandering teammate Frank who came up with the winning idea. "Take her to our rugby game this weekend, dude. Let her see you in action." Researchers have shown that nothing serves as a better aphrodisiac for women than a show of dominance and strength.

Nicole had never been to a rugby game and was surprised by how rough it was. She loved seeing Ryan steal the ball and take it downfield for the winning point, and Ryan loved knowing she was in the stands watching. She couldn't believe how turned on she was by his sweaty body. After the game, he glowed under the approving looks she was sending him and basked in the envy he saw on his teammates' faces. Ryan was pleased that this strategy seemed to be working.

While humans and animals have differences in their mating strategies, scientists have observed some curious similarities. One of the most colorful examples of animal tactics is provided by the side-blotched lizard (*Uta stansburiana*). Conveniently, the males come with three different colored throats that match their mating styles. Males with orange throats use the alpha-male harem strategy. They guard a group of females and mate with all of them. The males with yellow throats are called sneakers because they slip into the harem of the orange throat and mate with his females whenever they can get away with it. The males with brilliant blue throats—my personal favorites—use the one-and-only strategy. They mate with one female and guard her 24/7. From a biological perspective, the approaches of the orange-throated harem leader, the yellow-throated sneaker, and the blue-throated one-female type are all successful mating strategies for lizards and for human males, too. I affectionately call my husband a blue-throat.

THE HORMONE OF MONOGAMY

So, women may be anxiously asking, "How can I pick a blue-throat?" We have no surefire answer yet on what makes for a monogamous human male mate, but research on furry little mammals called voles might provide some clues. Scientists have found that male prairie voles are monogamous and share equally in parenting their offspring. But their cousins—the montane voles—are strictly promiscuous, seek sexual variety, and specialize in one-night stands that last less than a minute. The difference between the mating strategies of these vole cousins originates in the brain. When the prairie vole finds his partner, he mates with her over and over in a twenty-four-hour sexual marathon. This sexual activity changes his brain forever. An area of his brain called the AH—the anterior hypothalamus—memorizes his partner's smell and touch, leading him to aggressively reject all other females. This blissful day in the new vole couple's relationship is not only unforgettable, but biologically necessary. Memorizing her and thus merging the so-called love and lust circuits in his brain will initiate a lifelong preference for this one female.

During sex, both prairie and montane voles release vasopressin and dopamine, but only the prairie vole has the type of vasopressin receptors in his brain needed to make him monogamous. And when scientists experimentally blocked these monogamy-inducing vasopressin receptors in the prairie voles' brains, they didn't bond with their sexual partners. The love and lust circuits in their brains couldn't merge. What makes the difference between the vasopressin receptors in the prairie vole brain and the montane vole brain is their differing genes. The monogamous vole's vasopressin receptor gene is a longer version, and the promiscuous vole's is a shorter version. When scientists inserted the long version of the gene into the promiscuous montane vole, he, too, became monogamous.

Although the brain biology in men may turn out to be more complicated than it is in voles, humans have this vasopressin receptor gene too. Some men have the long version, while others have the short one. A study in Sweden found that men with the long version of the vasopressin receptor gene were twice as likely to leave bachelorhood behind and commit to one woman for life. So when it comes to fidelity, the joke among female scientists is that "longer *is* better," at least when it comes to the length of the vasopressin receptor gene.

CONFIRMED BACHELORS

Ryan's friend Frank was a confirmed bachelor and a master of seduction. He had well-rehearsed lines and knew exactly what to say and do to score sexual points. Phrases like "You're so beautiful you should be a model" and "I've never met a woman like you before" may be clichés, but Frank was so gorgeous and charming that women were ready to believe his little deceits. And according to studies, Frank has plenty of company in this "no strings, no commitment" mating strategy. They showed that deception serves as an important part of men's mating strategy for short-term partners. And researchers found that three out of four men said they were willing to lie or "modify the truth" to persuade women to have sex with them. They found that the things dating men lie about are similar around the world. Men exaggerate their wealth, status, and business and social connections. Frank frequently exaggerated his income and financial prospects and rarely missed a chance at name-dropping.

Now that Nicole and Ryan were seeing each other a few times a week, she sometimes heard Frank bragging about his latest conquests. She was so put off by his maneuvering that she tried to warn his newest girlfriend, Stacey. But it was no use. Stacey's brain and body had already fallen under Frank's seductive spell. What she didn't know was that each time she and Frank had sex, she was falling a little more in love—the oxytocin released during her orgasms was binding her body and brain closer to Frank. But it was working the other way around for him. He was starting to get bored. The more Stacey tried to pin him down to future plans, the more he squirmed. He felt it was time to move on. No cage of domesticity for him.

To be sure, both men and women try to manipulate the mating game. But when it comes to using verbal deception, researchers have found that men are biologically more comfortable with it than women. They measured the vocal strain of men and women telling lies to the opposite sex and found that the men showed much less electrical strain while they lied. This allowed the men like Frank to deceive in a more convincing manner. Ryan was glad Nicole didn't meet Frank first, because he knew Frank would have made a play for her immediately. She didn't seem like the type to fall for Frank's short-term mating style, but he'd been wrong about women before.

THE MALE BRAIN IN LOVE

When Nicole finally invited Ryan to spend the night, he felt like he'd died and gone to heaven. After that, they made love every day, sometimes more than once, and he still couldn't get enough of her. Sex doesn't always lead to love, but for the male brain, it is a necessary part of getting there.

Ryan's brain on sex was producing chemicals that create a blissful euphoria, similar to being high on cocaine. He couldn't figure out why, when he was away from Nicole for more than four or five hours, he started getting a primitive biological craving. If we could travel along Ryan's brain circuits on a miniature train as he was falling in love, we'd begin in an area deep at the center of his brain called the VTA, the ventral tegmental area. We'd see the cells in this area rapidly manufacturing dopamine—the brain's feel-good neurotransmitter for motivation and reward. As the train was being filled with dopamine at this VTA station, Ryan was starting to feel a pleasant buzz.

Filled with dopamine, the train would speed along his brain circuits to the next station, the NAc, or nucleus accumbens, the area for anticipation of pleasure and reward. Because Ryan is male, we'd see the dopamine from the train being mixed with testosterone and vasopressin. If you're female, it gets mixed with estrogen and oxytocin. Mixing dopamine with these other hormones was now making an addictive, high-octane fuel, leaving Ryan exhilarated and head over heels in love. The more Ryan and Nicole made love, the more addicted their bodies and brains became.

When the lovebirds were apart, they were constantly thinking about and texting each other. The love train with its addicting fuel makes it so we can't stop thinking, fantasizing, and talking about the person we're in love with. In one study, men and women said they spent up to 85 percent of their waking moments daydreaming about their lover. Ryan felt as if he were literally incorporating Nicole's essence into the fabric of his brain circuits. And he was. As the train sped into the final station, the caudate nucleus, or CN, the area for memorizing the look and identity of whoever is giving you pleasure, we'd see all the tiniest details about Nicole being indelibly chiseled into his permanent memory. She was now literally unforgettable. Once the love train had made these three stops at the VTA, the NAc and the CN, we'd see Ryan's lust and love circuits merge as they focused only on Nicole.

MATE-GUARDING

Ryan was starting to think of Nicole as "the one," and he was determined to hold on to her—so much so, Nicole told me, that when other guys were around, Ryan was sure to take her hand or possessively put his arm tightly around her.

When she came in for her appointment, she said, "I like it that he's so protective, but it seems a little hypocritical. I mean, just a couple weeks ago I caught him checking out a girl with big boobs at the car wash."

She told me that Ryan had been putting the quarters in the coin slot when a minimally dressed twenty-something walked by. "You should have seen the look he gave her. I know guys look at other women, but I can't believe he did it right in front of me. Maybe he's more like Frank than I thought."

I told Nicole that the lust center in the male brain automatically directs men to notice and visually take in the details of attractive females. When they see one that lights up their sexual circuit board, their brain instantly produces a quick sexual thought, but then it's usually over. To Ryan's mating brain, the buxom woman was like a bright, colorful hummingbird. She flew into his line of vision, caught his attention for a few seconds, and then flew off and out of his mind. For many men, this can happen several times a day. Ryan couldn't have stopped his eyes from looking at her breasts even if he'd tried. But he could learn to be more discreet. Because this is an autopilot behavior for the male brain, men don't think it's a big deal, and they can't understand why women find it so threatening. Until the tables are turned.

Unbeknownst to Ryan, Frank was looking forward to an

upcoming rugby game for reasons that went beyond his need to beat their biggest rival. He was secretly excited about Ryan being out of town that weekend, leaving Nicole alone and "unguarded."

Not wanting to raise Ryan's suspicions, Frank waited until just an hour before the rugby game to casually text Nicole and ask, "Want a ride to the game?"

She was on her way home from work, and going to the game sounded like more fun than going home alone. "Sure," she texted back.

She got so involved in the game that she didn't text Ryan to tell him where she was until his team was set up for the winning point. And then she excitedly sent him a text: "You're winning."

At first Ryan was confused. So he texted her back, "Where are you?"

"At the game," came the reply.

Ryan paused, then texted, "With who?"

She momentarily hesitated and then wrote, "Frank gave me a ride."

Ryan's heart sank. He couldn't help becoming furious as he imagined Frank hitting on her. He pushed the autodial button for Nicole's cell number. But the roar of the crowd that followed the final point of the game was so loud that she didn't hear her phone ring.

It was clear to Ryan that Frank was sneakily trying to lure Nicole away from him—a tactic called mate-poaching. Men who mate-poach consider it a double victory: They beat the guy and get the girl. This scenario is well documented in the animal kingdom, too. Groups of primates such as chimpanzees can live happily together until a female is in heat and ready to mate. At that point, the dominant males will become rivals and fight for her attention.

When Ryan finally got through to Nicole, she was at the victory celebration. After just a few sentences, Frank snatched Nicole's cell phone and gave Ryan the play-by-play of the winning points. And then he hung up—but not before covertly turning off Nicole's ringer.

Ryan tried Nicole several more times before going to bed, and every time he got her voice mail, he got more worked up. He was ready to rip Frank's throat out. He'd never experienced such jealousy and rage. As he tried to fall asleep, he couldn't get the image of Frank kissing Nicole—or worse—out of his mind.

Researchers have found that the fear of loss or rejection can intensify our feelings of love. Ryan's mating hormones, testosterone and vasopressin, were igniting the fear-of-rejection center in his amygdala and the mating area in the hypothalamus. His brain was putting him on red alert and driving his territoriality and possessive mating instincts wild. As Ryan was discovering, deeply passionate feelings can lead to enduring commitments. By the time his plane landed the next morning, all he cared about was making Nicole his. He knew his heart wouldn't stop racing until he popped the question. Ring or no ring, he couldn't wait. He asked her that night and she said yes. There was no doubt that Ryan's love and lust circuits were now in sync. What would keep them in sync? Sex.

FOUR

The Brain Below the Belt

AFTER A messy divorce, my patient Matt was finally getting on with his life, and I was glad to know he was feeling good about himself again. A handsome thirty-four-year-old lawyer, he had first come to see me a few years earlier when his wife filed for divorce. At that time, he was twenty pounds overweight, and his self-esteem was plummeting. But over the past two years, Matt had worked through most of his anger, gotten back into shape, and regained his self-confidence. He had even started dating again.

I've witnessed this "rebound transformation" in many men. After a brief hiatus, Matt's brain biology was once again driving him to seek sex and encouraging him to pursue a variety of partners. Researchers have reported that men want an average of fourteen sexual partners in their lifetime, while the women said they wanted an average of one or two. Researchers surmise that some of the disparity in these numbers can be chalked up to men's interest in one-night stands.

Given Matt's brain reality, I wasn't surprised to hear that speed dating had become his favorite way to meet women. As scientists know, a man's testosterone rises when he pursues attractive women. And when Matt walked into a room full of speed-dating women, it made his testosterone rise even further. Researchers in the Netherlands found it took only five minutes of casually interacting with attractive women for men's testosterone levels to go up.

Six feet tall with dark wavy hair and deep brown eyes, Matt never had trouble attracting women. But like most men, he frequently wished he didn't have to make the first move, and speed dating eliminated that hurdle. When I asked Matt how he could tell if he wanted to date a woman based on a six-minute speed-dating meeting, he shrugged and said, "I just know." He said he could tell if he was sexually attracted to a woman before she sat down at his table or uttered a single word. Researchers at the University of California found that it takes the male brain only one fifth of a second to classify a woman as sexually hot—or not. This verdict is made *long before* a man's conscious thought processes can even engage. And often it's the brain below his belt that knows first.

MEASURING UP

The penis has always occupied a larger-than-life place in the minds of both men and women. But when it comes to sex, size is less important than men think. What many women don't know is that men can feel just as self-conscious about their bodies and genitalia as women feel. Being seen naked by a new partner isn't much easier for some of them than it is for us.

They worry about what we'll think of their bodies and can experience anxiety about the way their penis is shaped. And many men fear that their partners will find them too small and will be disappointed.

Even though most men say they wish they had a larger penis, 85 percent of women say they're happy with their partner's size. Women report being most turned on by other physical features, like his eyes, smile, jawline, and muscles. And when it comes to being selected as a long-term partner, studies show that men get more mileage out of improving their personality and their bank accounts than out of investing in penile enlargement. But regardless of what we women think, many men erroneously believe penis size is their most important feature.

The irony here is that most men have no reason to feel insecure about their penis size. The average penis is much larger than it needs to be. According to researchers in England, the average erect penis ranges from 5.5 to 6.2 inches. Compared with other male mammals, relative to their females, it's supersize.

THE AUTOPILOT PENIS

All men know that the penis has a will of its own and can rise to attention without a single command from his brain. These reflexive erections are different from true sexual arousal because they come from unconscious signals from his spinal cord and brain, not from a conscious desire to have sex. The testosterone receptors that live on the nerve cells in a man's spinal cord, testicles, penis, and brain are what activate his entire sexual network. Women are surprised that the penis can operate on autopilot and even more surprised that men don't always know when they're getting an erection. The autopilot penis is part of a man's daily reality for most of his life, though it happens less as he gets older. We women often notice the rising tide before he does.

True arousal for men typically starts in the brain with erotic thoughts or images. That's all it takes for a man's brain to send signals down the spinal cord to the penis to start an erection. As long as men have an adequate supply of testosterone, seeing an erotic image will automatically activate their brain's sex circuits. And as Dr. Frank Beach, my neurobiology professor at UC–Berkeley, taught us, "The male brain's sexual-pursuit and arousal circuits must be primed for action by testosterone in order for him to function."

This hormone increases sexual interest and revs up the horsepower of his thrusting muscles and penis for high performance. So, prior to their forties, seeing is often all it takes for most men to become fully erect. After that age, the frequency of the instant hard-on is reduced, and men often need some physical stimulation to become erect enough for penetration. Because Matt was in his early thirties, the connection between his eyes and his penis was more than adequate.

If we could observe Matt's brain with a miniature PET scanner on his weekend date, we'd see how it was directing the show. When his date walked out of her apartment and he laid eyes on her curvy figure draped in a slinky red dress, we'd have seen his visual cortex send a message to his hypothalamus to start the hormonal engines for erection. One look at her long legs in high spike heels got his complete attention. Her full lips and the blush of her cheeks registered as signs of fertility in his brain. As she tossed back her shiny brown hair and gave him that come-hither smile, he knew she might be sexually interested. This would light up his anticipation-of-pleasure center, the NAc (nucleus accumbens), letting him know there was hope of a sexual reward.

Sexual arousal begins in the brain, but it is reinforced by physical contact. Once a man is aroused, a woman's simple touch can send sexual tremors through his brain and body down into his penis. Later that evening, when Matt's date began unzipping his jeans, our miniature camera in his brain would show his hypothalamus directing blood to rush into his penis. We'd see a jolt of activation lighting up his frontal lobe's "pay total attention to this now" circuits. Matt's brain and body were now on high alert for taking advantage of this sexual opportunity. The instant his date gave him the green light, he'd be ready to hit the accelerator and head for the promised land. He was relieved that it didn't take long. Soon, *she* was pushing her hips up against his. With our camera, we'd now see all his brain areas *not* needed for sex going dark and deactivating. Any and all distractions were being silenced as his brain sent out the message *Penetrate now!* The urgency quickly spread throughout his body as he sucked in a quick breath of air. Ready, get set, go, and with one smooth thrust, he was inside.

To heighten his sexual arousal Matt played erotic fantasies in his visual circuits. These sexual fantasies had always helped him maintain a firm erection whether during sex or masturbation. Like most sexually active men, Matt had amassed a sex-fantasy DVD library in his brain. In his case, it was an erotic fascination with breasts that did the trick. Matt wasn't consciously conjuring up these fantasies. They were playing out on their own internal circuits, building sexual tension, arousal, and pleasure. Visual stimulation—even in fantasy—is what turns a man on, makes his penis hard, and keeps it up.

According to sex researchers, men are not only more stirred by sexual visions than women are, but also want to be more sexually adventurous. In a large national study, Dr. Edward Laumann and his colleagues counted and categorized men's and women's sexual acts like vaginal sex, oral sex, and anal sex. They found that across the board, it was men, not women, who suggested broadening their sexual repertoire. For example, men reported wanting to have group sex thirteen times more than women reported wanting to have it, and fellatio twice as often as women.

Oral sex can be a source of conflict in many relationships. Sex researchers believe that men enjoy this sexual act for many reasons, but one of the biggest is heightened sensitivity: the tongue, lips, and fingers can stimulate and stretch a man's urethral opening, increasing sensitivity in a way that doesn't happen as much inside the vagina. Researchers at McGill University found that as a man becomes more and more sexually aroused, the tip area, the glans of his penis, becomes less and less sensitive. This may be Mother Nature's way of protecting a man from pain during sexual intercourse. So, if there are times when a man can't reach climax during intercourse, he often *can* orgasm with the extra stimulation of oral sex.

ORGASM

In order to reach orgasm, both men and women must first turn off a few parts of the brain—like the amygdala, the brain's danger and alert center—and the areas for self-consciousness and worrying—the anterior cingulate cortex, or ACC. Aside from that, for men, arousal and orgasm are relatively simple and mostly about hydraulics; they need blood to rush to one crucial appendage. But for women, to turn off the worrying part of her brain, the neurochemical stars need to align. It takes more for a woman to get in the mood, relax, and deactivate her amygdala. That's why many sex therapists say that for women, foreplay is everything that happens in the twenty-four hours preceding intercourse, while for men, it's what happens three minutes before entry. But once men and women reach orgasm, the differences are few. Researchers have studied men and women in a PET scanner while their partners were manually stimulating their penis or clitoris to orgasm. Although differences showed up between men and women while they were being stimulated, there were few, if any, discernible brain differences during orgasm itself.

When Matt reached the point of no return, his brain released all the brakes and he emitted a deep, involuntary groan. As he climaxed, his brain circuits and body were flooded by norepinephrine, dopamine, and oxytocin, increasing his ecstasy. Simultaneously, his brain area for intense pleasure, the ventral tegmental area (VTA), and his brain area for pain suppression and vocalization—the periaqueductal gray (PAG)—activated intensely. Tonight, Matt's timing was perfect, and he felt his partner's vagina contracting with the waves of her orgasm at the same time that he had his—intensifying the pleasure for both of them.

Until men learn to inhibit the sex-arousal centers in their brains, the tail wags the dog, and they often reach orgasm long before their female partners have a chance. For reasons scientists don't completely understand, it typically takes women seven to eighteen minutes of vaginal intercourse to climax, and Matt was pleased that he'd already mastered the self-control problems of his early twenties. Scientists have discovered a group of neurons in the spinal cord called spinal ejaculatory generators that can be turned on or off by the brain. To gain dominion over the brain below his belt, a man must learn to direct his focus from his brain's sex centers to a nonsexual area. Tricks men may use to accomplish this include mentally solving complicated math problems, silently reciting the alphabet backward, or activating the disgust center, the insula, by thinking of something revolting. But when his penis is being pumped up with ten times the normal amount of blood, trying to stop an orgasm can be like trying to stop a runaway train. Perhaps that's why up to 40 percent of young men climax in fewer than eight to fifteen penile thrusts. According to researchers, more experienced men like Matt can teach themselves to last for seven to thirteen minutes or more.

PERFORMANCE ANXIETY

Matt was glad that he finally had some staying power and that his days of hair-trigger ejaculating were well behind him. A condition called premature ejaculation, or PME, can be a source of embarrassment for men and frustrating for them and their partners. Also known as rapid ejaculation, it affects between 25 and 40 percent of men in the United States, and most men have experienced it at least once. Aside from a lack of physical control, it can be caused by a variety of psychological factors, such as stress, depression, a history of sexual repression, and unrealistic expectations fostered by the media about men's performance.

Men who have high expectations about their sexual performance can sometimes experience an inability to become erect or stay erect long enough to have sexual intercourse. When Matt came in to see me about having this sort of trouble, he told me that he'd never experienced it until recently, and he was worried that something might be wrong with him. After dating a variety of women for several months and having no sexual difficulty whatsoever, Matt met a woman named Sarah, whom he was more into than any other woman he'd ever met. Sarah was a twenty-nine-year-old dancer with a gorgeous face and a body to match. He said, "I didn't want her to think I was just after her body, so we went out a couple times before I even made a move." He wanted her to trust him and feel comfortable before they had sex because he knew she'd need to be relaxed to have an orgasm. And he was determined to give her the best one she'd ever had.

But he was horrified to find that he'd put so much pressure on himself that when the time finally came to have sex, he could only get semierect. He was afraid she'd take this as a sign that he wasn't that into her, when in fact the opposite was true. Many men have told me that their "stage fright" is proportional to how hot the woman is and how much they want to impress her. This is when the so-called simple hydraulics of a man's sexual system can stall.

As a matter of fact, Matt's worrywart and performance-anxiety center, the ACC, was shutting down his spinal generators for erection, as well as his capacity to relax. This meant he couldn't stay hard enough to penetrate. Deep in his brain, his amygdala and ACC were triggering the fight-or-flight system in his sympathetic nervous system (SNS) and thereby turning off the neurochemicals in his hypothalamus and parasympathetic nervous system (PNS) that are needed for an erection. The PNS induces the chemical relaxation response that allows the blood vessels in the penis to open up and fill with blood, thus achieving an erection. When a man feels relaxed, his brain's PNS and oxytocin cells release oxytocin down the spinal nerves to aid in penile erection. So the correct balance between the PNS and SNS is crucial for a man to get an erection.

Matt was very worried that if this happened to him again, Sarah would think something was wrong with him and his chances for an ongoing relationship with her would be shot. He wanted me to prescribe a Viagra-like drug as "insurance" so he could be sure to get the erection he wanted. (Drugs like Viagra keep blood inside the penis to get and maintain an erection.) He said, "I've read about guys who had this performance problem happen, and once it happens again, it can become 'a thing' and keep happening." Matt was right. A man's anxiety over a previous failure can lead to more failures. This sort of performance anxiety can happen to men at any age, but because Matt was in his early thirties, I thought he could probably eliminate his difficulty by being more relaxed before he had sex with Sarah again. An intense physical workout before a date can sometimes do the trick, and some studies show that one or two alcoholic drinks may aid the relaxation response too. "Don't have more than two drinks, though," I warned him, "since as you probably already know, too much alcohol can make erection almost impossible."

POSTCOITAL NARCOLEPSY

The next time Matt came in to see me, he was feeling good about himself. He said the combination of a two-mile run and a couple of beers had worked for him, and things were going very well with Sarah. But after a couple of minutes of bringing me up to date on how things were going, he asked, "One other thing I wanted to ask you is whether it's normal or not to fall asleep right after sex."

I told him that this is something nearly all women complain about; they think it's a sign that the man doesn't care enough about them to stay awake and cuddle for a while. But the truth is that the hormone oxytocin is to blame for a man's so-called postcoital narcolepsy. Oxytocin promotes pleasurable, warm, safe feelings during and after sex for both men and women. In the female brain, the oxytocin and dopamine released after orgasm make her want to cuddle and talk. But research shows that this postorgasmic blast in men may lull them to sleep as it's released into their hypothalamus, triggering the brain's sleep center. I said, "For reasons we don't yet understand, in men it works a lot like a sleeping pill."

Indeed, it turned out that Sarah *had* felt neglected when he fell asleep after sex. However, he wanted to do everything he could to keep her in his life. He told me, "There's some truth to the stereotype that guys care most about sex, sports, and beer, but a lot of us want a long-term relationship, too. I have a good connection with Sarah, and even though we've only been going out a couple months, I think things could end up getting serious."

I was pleased that Matt had regained the courage to consider a long-term relationship again. He'd mentioned more than once that he'd like to have a family someday, and I had a suspicion that if things continued to go well with Sarah, she might end up being his wife and even the mother of his children.

FIVE

The Daddy Brain

OH, SHIT! *This can't be happening,* Tim thought to himself as Michelle showed him the bright pink line on the home pregnancy test. Tim, a muscular thirty-four-year-old contractor, had that deer-in-the-headlights look as he tried to hide his panic from his wife. They'd been married for just six months, and although Tim wanted kids, it was too soon. In their initial couples-therapy session, they told me they were planning to wait a few years before starting their family. This wasn't following their blueprint. Now the words his older brother Mike had said were haunting him: "Fatherhood changes your life forever, dude."

Mike was right. Some men are over the moon about their wife's pregnancies, but studies show that feelings of distress peak for most men four to six weeks after they discover they're going to be fathers. They seldom reveal these worries to their mates, and Tim's way to handle anxiety about being a father turned out to be by arming himself with information. He asked me to suggest some books on pregnancy, childbirth, and parenting. He also went online for information, and it turned out that some of what he read just raised more fears. For example: "The way parents tend to a baby's needs during the first weeks and months after birth can shape the baby's brain for the rest of its life." By the time Michelle had her three-month prenatal visit, Tim's nerves were *more* jangled, not less.

The turn for the better came when Michelle lay on the table for her first ultrasound. Tim was sitting alongside her as the doctor rubbed cold gel on her belly and turned on the machine. When a baby's image appeared on the screen, Tim audibly gasped as he saw its heart beating. "It was like nothing else mattered," he said. "All I could do was stare at this tiny beating heart and think, 'Oh my god, that's *my* child.' "

Scientists now know that a man's brain changes as his mate's pregnancy progresses. Dads typically don't crave pickles with ice cream or wake up nauseated every morning as moms do, but they do have emotional, physical, and hormonal shifts in parallel with their mates' pregnancies. Research at Harvard University revealed that two major hormones change in fathers-to-be: testosterone goes down and prolactin goes up. Scientists believe that men may be responding to the natural airborne chemicals of pregnancy—pheromones—emanating from the mother-to-be's skin and sweat glands. Unbeknownst to him, these hormones are priming him for paternal behavior. In some men, this hormonal shift can cause couvade syndrome—"sympathetic pregnancy." Couvade syndrome has been documented in fathers-to-be worldwide, and Tim was experiencing it firsthand. By the beginning of Michelle's second trimester, she needed bigger clothes—and so did Tim. He'd gained fifteen pounds.

And in a biological tit for tat, at least in mice, the father's pheromones have been found to waft through the air and into the mother's nose and trigger her to make more prolactin, a hormone that increases the growth of maternal brain circuits. The mommy brain begets the daddy brain, and the daddy brain abets the mommy brain.

As Michelle's belly and due date loomed large, she spent hours refolding tiny baby clothes and blankets and collecting all the other baby supplies she thought they'd need. Meanwhile, Tim was also "nesting." He became obsessed with fixing up the house, painting the baby's room and building shelves for all the new infant equipment, books, and toys. Scientists have found that men have the biggest hormonal leap from non-dad to dad in the days leading up to the birth. Researchers studied fathers-to-be during the last trimester of their wife's pregnancy and found that these men's prolactin levels increased by over 20 percent and their testosterone dropped 33 percent during the three weeks before birth. And by the time their children were born, not only had the fathers' testosterone dropped, but they were better at hearing and emotionally responding to crying babies than non-dads were. On average, a man's testosterone and prolactin levels will begin to readjust when the baby is six weeks old, returning to prefatherhood levels by the time the baby is walking.

In cultures around the world there is a lot of variability among fathers. Dads who are actively involved in taking care of their children have been found to have lower levels of testosterone than uninvolved fathers. Researchers compared two different hunter-gatherer cultures, one in which fathers give a lot of hands-on care and another in which fathers give very little care. Hadza dads, who give more hands-on care, had lower testosterone levels than the dads in the Datoga tribe. Datoga dads have very little contact with their children, and they had higher testosterone, closer to the levels of the single men in that tribe. No one knows for sure whether the different hormone levels cause the behavior or whether the hands-on fathering suppresses testosterone.

A DAD IS MADE

A week after Michelle's due date, Tim rushed her to the hospital as her contractions intensified. For the next thirty-six hours, he stayed awake, helping her to breathe through the contractions and trying to make her feel more comfortable, which seemed like an impossible task. During the birth itself, Tim couldn't believe how hard Michelle had to work. He was never so glad to be a man. Twice he felt he might faint. And then suddenly he could see the crown of the baby's head, and he became completely transfixed as the entire head and shoulders began to emerge. When the doctor handed Tim his newborn son, tears welled up in his eyes as he snuggled naked little Blake against the bare skin of his chest and neck.

"When he looked into my eyes, I think he knew I was his dad and I would always protect him," Tim later told me. The skin-to-skin contact between father and son had worked its biological charms on both of them, calming them and promoting bonding.

Because infants require round-the-clock care for survival, Mother Nature has forged a nearly unbreakable biological bond between parent and child. It's as if she waves her magic wand over the parents' brains and they fall head over heels in love with their baby, as Tim and Michelle were discovering. Scientists have learned that the same brain circuits that were activated when Tim and Michelle fell in love were now being hijacked to make sure they fell in love with Blake. Cupid's arrows were being dipped in powerful neurochemicals like dopamine and oxytocin. Just as in romantic love, the connections between the baby's and parents' brain circuits are reinforced by skin-to-skin contact and gazing into each other's eyes and faces. And researchers have shown that a baby's face, with its soft, pudgy cheeks and large eyes, activates a special brain area called the parental-instinct area within a seventh of a second. Tim and Michelle's instincts were turned on full blast.

DAD'S TENDING INSTINCT

"That little fellow sure has a good set of lungs," Blake's grandfather said as his daughter swooped into the room and popped the pacifier back into Blake's wailing mouth. Crying is a universal caretaking cue, but it stimulates the brains of fathers and mothers differently. Fathers' and mothers' brains light up in similar areas when they hear a baby cry. But the mommy brain activates more intensely, which may be why she's compelled to stop the crying before the dad feels compelled. So when Blake cried, Michelle almost always got to him first, even if Tim was closer. He was astonished by how quickly she heard and responded to their son's every whimper. But Tim's tending instinct and response to Blake's cries were improving daily.

As it turns out, the tending instinct is prewired into all human brains, not just mothers'. If we could have taken a brain-scan camera inside Tim's head as he cared for Blake, we'd have seen his amygdala, his worrywart ACC, and his insula—the area for gut feelings—light up as he heard Blake crying. Then, as Tim playfully changed Blake's diaper and kissed his soft stomach, the gleeful smile on his son's face would trigger his brain's reward center, the NAc, or nucleus accumbens. At this moment, all the circuits of Tim's daddy brain would be pulsing with the joy of fatherhood. Tim's brain was being stimulated to make new connections to reinforce his tending instinct. And each new connection in his brain helped him to get more in sync with his son.

FATHER-INFANT SYNCHRONY

New fathers are often surprised by how much they want to hold and play with their babies. The making of a daddy brain requires not only hormones and paternal brain circuits but also physical touch. At Princeton University, researchers compared dads and non-dads in our primate cousins the marmosets. Marmoset dads are probably the most involved fathers on the planet, holding their newborns more than fifteen hours a day, every day, for the first month. Carrying the infants for so many hours each day aligns the dads' brains with their offspring. The researchers found that the area of the marmoset dads' brains for thinking and predicting consequences, the prefrontal cortex (PFC), showed more cells and connections than in the non-dads. This brain area has receptors for the so-called fathering hormones: prolactin, oxytocin, and vasopressin. These scientists concluded that the experience of being a hands-on father dramatically increases the number of connections in the male brain for paternal behavior. Brain-scan studies show that contact between parent and child also activates the PFC in humans. So even though moms' brains may be on higher alert from day one, it's now clear that dads' brains can quickly catch up. Tim didn't need a brain scan to tell him what he already knew—that the same brain that used to be glued to Sunday football was now completely absorbed with Blake.

Because Tim had been involved from the day Blake was born, his daddy brain circuits were now running like a well-oiled machine. Even though Blake couldn't talk, he and Tim had been establishing an understanding and getting to know each other. Researchers' technical word for this parent-child understanding is *synchrony.* Synchrony is like an extended series of volleys in a tennis match. Some examples are tickling, eye contact, laughter, and teasing. This back-and-forth interaction in games like peekaboo is critical for developing parental behavior, according to studies by Dr. Ruth Feldman. Many fathers who don't have daily hands-on contact may fail to form the strong daddy brain circuits required for parent-child synchrony. The environment for eventually establishing such a close interaction may start before birth. During the last months of my pregnancy, my son's father would play a tapping game with him. His dad would *tap tap tap* on my belly, and he'd *tap tap tap* back—kicking seemingly with the same rhythm. The father-son relationship had begun.

DADDY AND MOMMY ARE DIFFERENT

Soon after birth, a baby can tell the difference between Mommy and Daddy. Within weeks of being born, Blake could see and smell the difference between Michelle and Tim, and he could hear and feel the differences too. Daddy had a deeper voice. Mommy had softer hands and talked as if she were singing. Even in the dark of night, Blake knew which parent was bending over his crib to take care of him. But Tim confessed that he couldn't help feeling a little jealous that Blake often seemed to want his mom more—and Michelle seemed to sometimes prefer Blake to Tim, too. As Tim was discovering, for fathers, early on, it's hard to match the biological force of the love bond between Mommy and baby. The baby is initially more bonded with the parent who has the yummy milk-filled breasts, and the intense pleasurable sensations of breast-feeding reinforce the mother's bond with her baby.

Scientists believe that the emotion and communication centers in the baby's brain learn to relate differently to Mom and Dad. This doesn't become obvious to the parents until the baby is about three months old and begins spending less time sleeping and more time interacting. At this age, Dad begins to play a starring role in baby's life. By the time Blake was six months old, he loved the stimulating games Tim played with him. When Tim kissed Blake's belly and blew loudly against his skin, tickling him, they were in their own private world.

ALONE TIME WITH DAD

Research shows that dads behave differently with their babies not only when the moms are away but also when the moms aren't watching. And infants notice the difference, too. One study showed that when Mommy, Daddy, and baby were all together, there were fewer interactions between Daddy and baby. And when fathers were alone with their babies, their playtime was much more spontaneous.

Establishing this comfortable spontaneity requires spending one-on-one time together, but some fathers, like Tim's brother Mike, either don't take or don't have this opportunity. Tim said the last time he stopped in at Mike's, he saw Mike's wife, Cynthia, snatch their eight-month-old son, Nathan, from his arms when the baby began to whine. Mike had been complaining for months that Cynthia didn't trust him, often criticizing or correcting his parenting. Tim said Mike had been looking forward to being a dad and imagined himself having fun with Nathan, but apparently Cynthia had a different plan. The only time she handed the baby to Mike was when her mother wasn't there. Then she'd thrust Nathan into Mike's arms the instant he walked in the door after work. Researchers at Ohio State University found evidence that fathers' beliefs about how involved they should be in child care didn't matter; it was mothers who were in the driver's seat. They discovered that moms are the gatekeepers for fathers' access to their children. Mothers can be very encouraging to fathers and open the gate to their involvement, or they can be critical and close the gate. Cynthia was a negative gatekeeper and didn't trust anyone with Nathan except herself and her mother. She was unknowingly operating on ancient brain circuitry that told her that female kin were the ones she could look to for help.

Whereas many fathers feel overwhelmed by being the family's main provider and also being expected to help equally with child care, Mike was begging for more time with his baby and couldn't get it. Tim was grateful that Michelle trusted him with Blake and didn't expect him to fill the "ancient shoes" of all her female kin. Michelle was also strengthening her marriage by letting Tim be a dad. Researchers have found that moms who are the least critical of their husbands and encourage the dad's interactions with the child fare best when it comes to staying together.

DADDY TIME IMPROVES SELF-CONFIDENCE

At twelve months, Blake climbed all over Tim as if on a human jungle gym and was constantly trying to pull his daddy to the floor so he could wrestle with him. When he succeeded, he'd triumphantly sit on Tim's chest grabbing his dad's face hard with his tiny hands and squeezing his chin or cheeks. Even at this age, Blake loved to test his skills with his daddy, and he loved it when Tim swung him into the air and back down again, trying to snag Tim's sunglasses or hair whenever he got close enough. Father and son playfully challenged each other at every turn.

Researchers have shown that the particular way fathers play with their children makes their kids more curious and improves their ability to learn. Compared with mothers', fathers' play is more physical and boisterous. The researchers found that daddy play is more creative and unpredictable and thus more stimulating. Fathers' creativity shows up not only while playing but also while talking and singing to their children. Researchers at the University of Toronto found that mothers sang the correct versions of "Twinkle, Twinkle, Little Star" or "The Itsy-Bitsy Spider," and fathers altered the verses, creating complex songs with unpredictable endings. The dads were more quirky and fun.

And that's not the only difference. In another study, in Germany, scientists followed a group of children for fifteen years. They first began observing the fathers interacting with their children at two years of age. They found that the children whose fathers played roughly with them, the way Tim played with Blake, were the most self-confident by the time they reached adolescence.

TEASING: THE SPIRIT OF MALE COMMUNICATION

Physical and verbal teasing is the way dads connect with their kids. Michelle wasn't thrilled when, thanks to Tim, five-year-old Blake's favorite phrase had become "You're a poo-poo head," which he would often say while vigorously pointing to his rear end. But for Blake and Tim, it was just part of their fun.

Dads employ teasing with sons and daughters, but their daughters usually don't like it as much as their sons. Daughters will soon try to divert Dad and assign him a part in the role-playing games girls prefer. (And most dads are usually willing to go along with whatever roles their little girls give them.) Boys, on the other hand, love the teasing games and will actively egg Dad on, trying almost anything to get his goat. Researchers have found that this kind of father-child play improves children's ability to guess what's on another person's mind and to recognize mental tricks and deceits. And for sons, this playful teasing with Dad establishes the foundation for building close connections with other males later in life. By the time Blake was six, he and Tim could banter mock insults for longer than Michelle could stand to listen.

HARSH TALK PREPARES CHILDREN FOR THE REAL WORLD

The next time I saw Michelle, she was annoyed with Tim. "Sometimes he barks orders at Blake like he's a drill sergeant," she said. "He'll practically bellow, 'Sit down, don't move, be quiet.' Even though Blake doesn't seem to mind, I think Tim is being insensitive and rude." I had heard this many times before. To mothers, fathers can sound too harsh, and to fathers, mothers can sound too soft.

Researchers have indeed found that fathers give their children more direct orders than mothers do. And mothers stay emotionally in tune with their children more, so they don't need to give as many direct orders. Mothers use shorter sentences and match their children's tone of voice more than fathers do. Michelle was right—Tim's style *was* rougher than hers, but not better or worse. Researchers believe that daddy talk provides an important bridge to communicating in the real world, where children will soon find that others can't read their minds or anticipate every need the way Mom does.

DADS AND DISCIPLINE

Discipline was another area where Tim and Michelle differed in style. Tim thought it was his duty to guide Blake's development by using firm rules and strict corrections. And he was not alone.

Researchers have found that fathers in cultures around the world believe it's their job to make their children, especially sons, toe the line. Of course, fathers must walk a fine line between overly harsh punishment—which creates fear, distrust, and a desire for revenge—and "just right" discipline. And even though some modern parenting styles endorse the laid-back father as more likely to be a good dad than the high-testosterone macho man, biological research suggests that the opposite may be true.

Male parents across the animal kingdom are typically more strict than female parents and more aggressive in handling their offspring. Although we don't think of aggressive males as being better dads, for some fathers, like rodents, being a good dad requires being rough-and-ready. A rodent dad must be aggressive and forcefully grab and retrieve runaway pups, or they will die. As with human dads, the rodent dad's paternal brain circuits are fueled by the hormones testosterone and vasopressin. The researchers discovered that the most aggressive rodent dads had the highest levels of these hormones. And interestingly, they also found that the most aggressive male pups grew up not only to have the highest testosterone but also to be better, more protective dads.

Good dads can be both aggressively playful and aggressively protective. Tim smiled as he remembered his dad's overly aggressive protection at one of his peewee football games. "I got fouled by this huge kid, and Dad went after the kid's dad and wanted to take his head off. The coaches had a hard time calming them both down," he said, laughing.

And according to researchers in Sweden, active discipline from fathers can be a key factor in children's success. In the study, children of dads who were active disciplinarians (meaning that they were strict, not that they beat their children) achieved better grades and went further in school than children of dads who were not disciplinarians. The sons of dads who were disciplinarians had fewer behavioral problems, and their daughters had fewer emotional problems.

DADDY'S LITTLE GIRL STEALS HIS HEART

Research also shows that when a little girl has a close relationship with her dad, it sets the stage for getting along better with men later in life. When Tim visited his friend Zack and saw him softly stroking his four-and-a-half-year-old daughter Kelsey's hair, he was amazed by how gentle he'd become. Zack, the same guy who had pummeled countless grown men on the college football field, was now sitting down to have tea parties. Tim watched in wonder as Zack played the roles Kelsey assigned him, including being her horsey and letting her ride on his back while he crawled around on all fours.

Daughters are notorious for wrapping daddies around their little fingers, and Tim witnessed an example of this later that day when Kelsey scolded Zack for setting the tea party table incorrectly: "The spoons go on *this* side of the plates, Daddy. The cups go on the saucers, not on the tablecloth. And you're supposed to put your napkin on your lap. Now let's start over."

Zack did as he was told, and Tim couldn't help but laugh. He asked Kelsey, "Why do you like to have tea parties with your daddy when he does everything wrong?"

"Because he does what I tell him to do," she answered matter-of-factly. When Kelsey played with Mommy or other girls, there was always more negotiating and compromising.

When she accidentally broke one of the cups and burst into tears, Zack came to the rescue with the Super Glue and made everything okay again. A study in Wisconsin reported that fathers feel closest to their daughters when they are doing something to help them. This holds true whether the daughter is four or forty-four. Dads bond with their daughters by helping to solve their problems and fixing things that are broken, whether it's their dollies or their financial portfolios. Fathers also bond with their sons by helping them, but research shows that this "help" often centers on making the boys stronger and tougher. Studies show that dads feel it's their responsibility to toughen their sons up to be able to survive as a man in the real world. This sometimes leads them to inhibit displays of affection in favor of rougher handling. Even so, researchers find that not only do fathers identify with their sons, but sons look up to their dads as role models of what they're supposed to act like when they grow up.

LIKE FATHER, LIKE SON

Like Tim, fathers know it is their job to initiate their sons into the perilous world where boys become men. Tim's father was a hands-off kind of dad, and this was a role model Tim had decided not to follow. Although the traditional emotional structuring of the father-son relationship hinges on the authority of the father, Tim was determined to make his relationship with Blake about more than discipline. So far, he was succeeding. He played with Blake every day, gave him lots of hugs, and praised his accomplishments.

On the other hand, Tim knew that coddling his son would work to Blake's disadvantage, so he helped him to make right choices and made him do as much for himself as possible. When Tim took Blake hiking, Blake carried his own backpack and water just like Dad. Tim was proud that Blake wanted to imitate him, and he was determined to set a good example in every possible way. When Tim and Blake played follow-the-leader, Tim took turns leading and following so Blake could learn both roles well. And when he and Blake wrestled, he let Blake pin him at least once in every three matches. He also let Blake win other contests, like their father-son races and video games. Studies have shown that insecure fathers are unable to let their sons beat them at any game even when their sons are very young.

Tim was what researchers call a high-nurturing parent, and studies show that this type of parenting style is healthier for kids throughout life. The brain effects of high-nurturing and low-nurturing parents on college-age kids' brains showed that those who'd had low parental care in childhood ended up with hyperactive brain responses to stress, according to researchers. And these young adults released more of the stress hormone cortisol than peers who'd had high parental care in childhood.

And it's not just kids' brains that benefit from close physical contact. According to a study of fathers, close physical contact releases oxytocin and pleasure hormones in dads, too, bonding parent to child. One of Tim's favorite times with Blake was at the end of the day, after bath time, when he'd read story after story next to him in his little bed. Tim told me he especially loved it when Blake snuggled against him as he fell asleep. The more a man holds and cares for his child, the more connections his brain makes for paternal behavior. Tim's male brain had entered a new emotional reality. And oxytocin had helped his softer side to blossom, as it would throughout his manhood. The more both women and men know about how the daddy brain is formed, the more hope we have of turning our parenting partnerships into satisfying and supportive relationships and families. And this is just what the daddy brain needs to be at its best.

SIX

Manhood: The Emotional Lives of Men

WHEN I SAW Neil's name on my appointment calendar, I knew something was wrong. A forty-eight-year-old partner at a prestigious architecture firm, he and his wife, Danielle, had come to see me a few years earlier to discuss their teenage daughter. Neil was usually so levelheaded that his wife only half-jokingly accused him of being an emotionless android. But when I called him to find out what was up, his voice was cracking with emotion.

"Danielle slept in the guest room last night," he said. "Ever since she got promoted to manager, she comes home upset every day. I try to help her, but she gets mad at me and says I'm not being supportive enough and that I don't understand how she feels. I love her, but I can't handle all this emotional drama."

It is the classic complaint: Men accuse women of being too emotional and women accuse men of not being emotional enough. I hear it all the time in my office, and each thinks the other could just *decide* to be different—if he or she really wanted to. What they don't know is that the brain circuitry for emotional processing is different in men and women.

When Neil and Danielle arrived in my office, he was clenching his jaw, and she was dabbing her eyes with a tissue. "I've never been under so much stress in my life," she said as she dropped into the chair. "If my department doesn't get this season's inventory done on time, we'll lose thousands of dollars, and we're already in the red. I just want Neil to listen, give me a hug, and tell me how he knows I feel. But he goes into robot mode and starts telling me what I should do."

Neil shook his head and said, "That's not how I see it. I already told her I feel bad about all the pressure she's under. She wants me to listen to her and be sympathetic, but then she won't listen to my suggestions."

Neil had always been his firm's go-to guy for creative problem solving, so when Danielle wouldn't let him offer solutions, it was baffling to him. Anxiously pulling at his manicured beard, he said, "Seeing her cry and not being allowed to help her is torture to me."

The look on Danielle's tearstained face suggested that she thought Neil was exaggerating, but when women cry, it truly may evoke brain pain in men.

TWO EMOTIONAL SYSTEMS

Until recently, differences in how men and women feel and express emotions were thought to be due to upbringing alone. And to be sure, how our parents raise us can reinforce or suppress parts of our basic biology. But we now know that the emotional processing in the male and female brain is different. Research has suggested that our brains have two emotional systems that work simultaneously: the mirror-neuron system, or MNS, and the temporal-parietal junction system, or TPJ. Males seem to use one system more, and females seem to use the other system more.

If we could scan Neil's brain as Danielle complained about her problem and started to cry, we'd see both of his systems for reading emotions switching on. First, his MNS would activate. The mirror neurons that make up his MNS would allow him to briefly feel the same emotional pain he was seeing on Danielle's face. This is called emotional empathy. Next, we'd see his brain's analyze-and-fix-it circuits being activated by the TPJ as it searched his entire brain for solutions. This is called cognitive empathy. The male brain is able to use the TPJ starting in late childhood, but after puberty a male's reproductive hormones may cement a preference for it. Researchers have found that the TPJ keeps a firm boundary between emotions of the "self" and the "other." This prevents men's thought processes from being *infected* by other people's emotions, which strengthens their ability to cognitively and analytically find a solution.

As we watched Neil's brain formulating a solution, we'd see his cortex activate as he matter-of-factly asked Danielle, "How many people will it take to get it done on time?" Bristling, she'd shot him a hurt, resentful look and said, "What difference does that make? I have to figure out how to do it with the twelve people I have. You don't understand." Neil's brain would entirely miss the desperate tone of her voice, since his TPJ would be busy working out the solution and his MNS would no longer be activating. Thus he'd be unable to notice the hurt expression in her voice and on her face. Next, we'd see his TPJ and cortex flashing with excitement as he found the solution: "Hire temps. It'll cost less than the money you'll lose if you miss the deadline," he proudly declared. Then we'd see Neil's brain circuits for victory light up as feel-good hormones were released in response to finding an answer. But those lights would quickly dim as he saw Danielle burst into tears.

She was convinced that Neil's analytical response meant that he didn't understand how she felt and didn't care. But he did care. He was simply trapped in his male brain circuit loops. And she was trapped in her female brain circuit loops. His male brain was using the TPJ for cognitively processing emotions and getting to a "fix it fast" solution. Her female brain's MNS was misinterpreting his blank facial expression. The female brain uses the MNS to stay in sync with other people's feelings, so women are often put off by a blank face. Male or female, when we see an emotion on someone else's face, our MNS activates. The difference—for reasons scientists don't understand—is that the female brain stays in the MNS longer, while the male brain quickly switches over to the TPJ.

When Danielle told Neil her problem, for just a few milliseconds his face mirrored her expression, and in that fleeting instant, he did *feel* her distress. But his male brain isn't designed to wallow in anguish, so once it identifies an emotion, it quickly taps into the TPJ to complete the cognitive emotional processing. The male brain is like an express train: It doesn't stop until it reaches its final destination.

And if Danielle told her problem to her sister or one of her girlfriends, they would probably stay in their MNS's emotional empathy system and share her emotions. Although she had interpreted Neil's fast exit from the MNS as uncaring, he was actually trying to solve her problem and alleviate her emotional suffering. I could relate from similar experiences with my husband. He usually launches into his helpful solution without saying, "Honey, I know how you feel."

I turned to Danielle and said, "Neil uses the TPJ more than the MNS because the male brain is structured to seek solutions rather than continue to empathize. But that doesn't mean he doesn't care. Solving your problems *is* the way he tries to show his love and concern."

Neil smiled with recognition, but Danielle didn't look convinced. She said, "Well, but you'd never know he cares by the look on his face."

A GUY FACE

From childhood on, males learn that acting cool and hiding their fears are the unwritten laws of masculinity. And especially since his testosterone surge at age thirteen, Neil had been practicing his guy face to be sure he kept his emotions to himself. For a man to physically strike a pose of self-confidence and strength, he must train his facial muscles to mask his fear.

Because facial muscles are controlled by the brain's emotional circuits, scientists have been able to learn about emotions by measuring these muscles. Researchers in one study placed electrodes on men's and women's smiling muscle—the zygomaticus—and on the anger/scowling muscle—the corrugator. They recorded the muscles' electrical activity as the men and women viewed emotionally provocative photos. Much to the scientists' surprise, the men, after seeing an emotional face for just one fifth of a second—so briefly that it was still unconscious—were more emotionally reactive than the women. But it's what happened to the men's facial muscles next that helped me explain Neil's guy face to Danielle.

As the experiment proceeded, at 2.5 seconds, well into the range of conscious processing, the men's facial muscles became *less* emotionally responsive than the women's. The researchers concluded that the men consciously—or at least semiconsciously—suppressed showing their emotions on their faces. Meanwhile, the women's facial muscles became *more* emotionally responsive after 2.5 seconds. According to the researchers, this suggests that men have trained themselves, perhaps since childhood, to automatically turn off or disguise facial emotions. The females' expressions not only continued to mirror the emotion they were seeing on the face in the photo, but they automatically exaggerated it, from a grin to a big smile or from a subtle frown to a pout. They, too, had been practicing this since childhood.

Men's poker faces are one reason women tend to think they are "emotionally challenged." But as this study showed, it becomes automatic for men to keep their feelings to themselves.

THE HORMONES OF EMOTION

Because men use their TPJ more, they can't fathom why women want to spend so much time talking about their emotions and often getting more and more upset. I told Neil and Danielle that I once asked my scientist husband, "Why do men respond to emotional issues with logic instead of feelings?" He laughed and said, "The real question is why women don't."

Neil laughed too and said, "Now all I need to know is how to get Danielle to use her TPJ more."

What Neil and Danielle didn't know was that through hundreds of thousands of years, our male and female brain circuits have been fine-tuned to run on different hormones. In fact, our sex hormones might be partially responsible for our different emotional styles. Male circuits use more testosterone and vasopressin; female circuits use more estrogen and oxytocin. These hormones run certain brain areas—like the amygdala, hypothalamus, and even perhaps the MNS and TPJ—differently in men and women.

Scientists have been testing how men's and women's brains react when they're given the other sex's hormones. Researchers found that when men were given a single high dose of oxytocin (a hormone that females make more of), it increased their ability to resonate with other people's feelings. So when the men looked at photos of faces displaying subtle emotional cues, they read them more accurately. The scientists concluded that the men became temporarily more empathetic. In a separate study, researchers gave women a single high dose of testosterone and found that it temporarily made them more mentally focused.

What Danielle criticized as Neil's "unsupportive, unemotional robotic mode" was actually the result of his finely tuned TPJ, enhanced by his high testosterone. Because this state of mind is the male brain's daily reality, it's hard for men to believe that women don't see the world the same way as they do. But we don't.

Danielle turned to Neil and, only half jokingly, said, "Well, I wouldn't mind using my TPJ more as long as it doesn't give me a male ego!"

THE MALE EGO

Joe, a forty-five-year-old manager at a local car dealership, called me in great distress, saying his wife, Maria, my former patient, was going to leave him if he didn't come see me about his anger. He explained that she was furious with him for getting into a shouting match with a cab driver. "I'm not saying I'm proud of getting into it with him," he told me, "but I don't think it was such a big deal. The guy asked for it."

And for Joe, it wasn't a big deal. A man's brain area for suppressing anger, the septum, is smaller than it is in the female brain, so expressing anger is a more common response for men than it is for women. The anger-aggression circuits in the male brain are formed before he's born and get behaviorally reinforced during boyhood and hormonally reinforced during the teen years. And by adulthood, using these hormonally influenced circuits for social risk-taking and aggression have become a familiar part of his life. Men in their forties, like Joe, still have a lot of testosterone and vasopressin fueling their brain circuits, often giving them a hair trigger for anger. Studies have found that though men and women report that they *feel* anger for an equal number of minutes per day, men get physically aggressive twenty times more often than women do.

I'd barely greeted Joe and Maria in the waiting room when he affably launched into an explanation: "I want you to know that I've been working hard to keep my cool and be more sensitive to Maria, but sometimes I slip up."

Maria said, "That was more than a slip! I thought any second they'd start punching each other. Tell me, what would make a forty-five-year-old man act like that?"

I looked at Joe and asked, "Well, what *did* happen?"

He folded his arms across his chest as he said, "It was nothing. She just gets upset."

But if we could have watched Joe's brain circuits while they were stuck in bumper-to-bumper traffic behind the cab that was braking at every green light, we'd have seen Joe's anger-aggression circuits responding to his rising hormonal tide. As his frustration grew, we'd see his testosterone and stress hormone, cortisol, activating his amygdala and firing up his fighting circuits. When Joe flashed his lights at the taxi to speed up and the driver hit his brakes instead, we'd have seen Joe's motor cortex activate the muscles in his hands and arms as he banged on the steering wheel and blasted the horn. When the cab driver retaliated by slowing down more and braking erratically, we'd have seen Joe's brain being flooded with a mixture of adrenaline, cortisol, and testosterone. We'd have seen his "good judgment" circuits, the frontal lobes, go dark and offline as his right foot pressed down on the gas pedal to bumper-tap the cab—hard enough to splash Maria's coffee all over her dress.

As the cabbie slammed on his brakes and jumped out of his cab, we'd have seen Joe laser-focusing every cell in his brain and body for a fight. When Maria yelled, "Stop, Joe! What if he has a gun?" Joe's auditory system barely heard her. He'd already thrust open his door and was hurling his bulky frame out of the car.

Now, sitting in front of me, Joe looked as though he'd been sent to the principal's office for fighting on the playground. He knew he was in serious trouble with Maria, but he still thought she was overreacting. As for Maria, the encounter with the cab driver would be the straw that broke the camel's back. Looking down, she shook her head and said to me, "Someday his stupid male ego is going to get him killed."

Most men aren't proud of their knee-jerk reaction to being challenged, but as Joe put it, "It's just a guy thing."

I explained that Joe's male brain biologically saw the cab driver's actions as a challenge to his territory and dominance, and his brain responded with a series of chemical changes prompting his aggressive behavior. Looking at Maria, I said, "This brain biology doesn't give men permission to be uncivilized, but it does provide insight into why they defend their manhood so vigorously."

Addressing both of them, I said, "Basically, the hormone cocktail in the male brain is the underpinning of Joe's anger and aggression."

Joe unfolded his arms and leaned forward, saying, "I guess this hormone cocktail does get me in trouble, at least with Maria."

Smugly, I thought to myself, "We've made some good progress in our very first session." Boy, was I wrong.

AUTOCATALYTIC ANGER

Joe and Maria had been married twenty-two years, and from Joe's perspective, they had had a good marriage. Up until now, when she threatened divorce, he was on top of the world. They had a nice house, and even during the latest recession, he was the biggest earner at the dealership where he worked.

Although Maria was proud of Joe's success at work, she didn't agree with his opinion of their marriage. She had a mental tally sheet for every fight they'd ever had. Studies have found that men and women remember facts equally well, but women remember the details of emotional events better and longer. The brain has two independent memory systems. One is memory for unemotional objects or events, and the other is for memory enhanced by emotion. In emotional situations these two systems interact in important ways. Essentially, men remember facts and figures, but women record not only the facts, but also *every detail of the emotion that they're feeling.* So when Maria recalled a fight with Joe, she'd not only remember the facts, but she'd reexperience her sadness, anger, and fear all over again.

She said, "It doesn't take much to set him off. I walk around on eggshells waiting for him to blow. And then he follows me around the house from room to room shouting at me and getting more worked up."

Maria was describing a behavior that scientists call *autocatalytic,* or self-reinforcing, anger. Once some men's anger ignites, it's hard to stop. Their anger gets fueled by testosterone, vasopressin, and cortisol. These hormones reduce a man's physical fear of the opponent and activate his territorial fight reaction. When Maria yells back at Joe, his brain knows she isn't a real threat to him, so her anger just gets him more fired up. His anger is feeding on her anger and then back on his own. Scientists have found that when anger reaches the boiling point in some men, under conditions of high testosterone, it can produce pleasure, egging them on and making their anger harder to control. Joe couldn't admit it to me, because he almost didn't know it himself, but part of his brain was enjoying being angry and seeing her angry. He was getting a high from his anger.

This high was what Joe had been using for decades to win competitions. He knew from playing high-school football that getting angry got him fired up. And he now used that energy to help him win the sales contests at work. When men like Joe are in a competitive mood or looking for a fight, seeing their opponent get angry creates a strange sort of excitement. The intelligent part of the brain, our cortex, has learned to harness deep, primitive emotions—like anger—to our advantage.

According to studies, people prefer to feel emotions that are potentially useful, even if those feelings are unpleasant. Researchers showed that even though anger can cause flawed thinking by reducing the perceptions of risk and triggering aggression, anger can sometimes make us think more clearly. They concluded that anger prompts more careful and rational analysis of another person's reasoning, so in some instances, anger can make people *more* rational, not less. But while Joe's anger was paying off at work, Maria made it clear that it wasn't scoring any points at home.

She quietly said, "Joe, when you get mad, it always gets worse, and it really, really scares me."

At this Joe whistled through his teeth and raised his eyebrows. "But you know I'd never hurt you," he said, looking stricken. In cultures all over the world, men like Joe consider it perfectly acceptable to express their anger, especially when they feel they're being challenged. So men are often surprised to hear that their wives and children are actually afraid of them. Researchers have also found that high-testosterone men, like Joe, more than low-testosterone men, have a need to dominate others, and so they react more dramatically to being challenged. And this happens in the animal kingdom too. Studies in primates show that dominant males whose status is consistently challenged maintain higher levels of testosterone and are more aggressive than subordinate males. The higher the testosterone, the more invigorated and battle-ready the male brain feels.

When Maria glared at Joe or shouted back at him, she was unknowingly challenging his dominance, thus increasing his testosterone. She was causing the flames of Joe's anger to flare up, escalating and prolonging the fight.

"Okay, if that's how it works, I'll agree to stop glaring back," Maria said. "But he has to promise to walk away before he gets so mad that he can't shut up."

I looked questioningly at Joe, and he nodded. "All right. I'll try," he said. "But I just don't get why she's so upset about this all of a sudden. I've been like this my whole life!"

As it turned out, when Maria and Joe first started dating, she was flattered by his jealous, intense way of letting her know she was the one. She said she liked the way he aggressively stared down other guys he caught checking her out. In those days, Joe's tough-guy attitude made him more attractive to her.

Research shows that angry men get noticed more—not only by other men but also by women. Ironically, it was the same high-testosterone personality traits that had initially attracted Maria to Joe that were now driving them apart.

"But it isn't hopeless," I said. "The good news is that research shows that couples who argue have a better chance of staying together. The fighting you've been doing, while damaging in its own way, gives your marriage a better chance of survival than if you suppressed your anger altogether." It was clear to me that Joe and Maria still loved each other. I just needed to help them work on expressing their anger in less destructive ways. But Joe would still need his aggression to motivate him and maintain his place in the pecking order at work, as my patient Neil was finding out firsthand.

THE COMFORT OF STABLE HIERARCHY

The voice mail I received from Danielle had been left in my urgent box. She said, "Neil hasn't slept in days and I'm worried about him. Things at work are pretty bad right now."

As I soon found out, when the president of Neil's architecture firm announced that he was retiring, Neil felt as though he'd been punched in the gut. He and his boss had always worked well together, and Neil would not only miss their camaraderie, but also his backing in these fragile economic times. Even though Neil got along with Ben—the VP who would soon be the new president—he hadn't worked on many projects with him and didn't know him well. Until now, Neil's position had always seemed secure. But now there was an atmosphere of uncertainty as the pecking order was about to change. And Neil wasn't sure that the changes would be for the better.

At home, Neil, who typically had a rock-steady demeanor, had morphed into a touchy and irritable grouch. Lately, even when Danielle said something that would normally make him laugh, he'd scowl. This wasn't the Neil she knew. When he started tossing and turning all night, she begged him to come see me. When he came in, he said, "I have to keep as much control at work as I can, so I applied for the VP opening." But Neil wasn't the only applicant. Four of the other senior architects in the firm applied too—including Neil's biggest adversary, George.

Research has found that when men are in a stable hierarchy, their testosterone and cortisol are lower than when they're not, reducing their tendency toward anger and aggression. A male's tendency to violence can be either dialed up or dialed down by social conditions. Scientists have found that a stable social hierarchy and a stable marriage are two factors that dial it down.

At least Neil has the stable marriage part, I thought. But when the pecking order is thrown into question, as at Neil's firm, even calm men start pumping out more testosterone, cortisol, and vasopressin, preparing them for turf wars.

Neil said, "I was doing okay until I found out George was going after the VP promotion too. That's when I stopped sleeping."

If we could take a look inside Neil's brain as he reacted to this territorial challenge, we'd see bursts of testosterone, cortisol, and vasopressin flooding his circuits. As he tossed and turned in bed while thinking about how awful it would be if George became his superior, we'd see Neil's territorial fear circuits activating in his hypothalamus and amygdala. As he pounded his pillow into the right shape for the tenth time that night, his mind would be buzzing with ideas—all directed at beating George out of the job. These thoughts would be stimulating, instead of relaxing, the "sleep cells" in his suprachiasmatic nucleus, or SCN. Now Neil's eyes were wide open as he ruminated about the hierarchy at work.

Research shows that social hierarchies guide behavior in many species, including humans. The mental machinery for jockeying for position is wired into the male brain. Fierce male-male competition is found in animals as diverse as lizards, leopards, and elephants and it's ubiquitous in higher primates.

Like human males, chimps will bluff, scheme, and even murder to gain or maintain rank. And like human males, they respond biologically to victories and setbacks. The testosterone that runs their competitive circuits ramps up as they anticipate a confrontation. Neil's brain was instinctively preparing him to do battle.

Evolutionary biologists suggest that behaviors like bluffing, posturing, and fighting have evolved to protect males, especially from opponents within their own species. Instinctive male-male competition and hierarchical fighting is driven by both hormones and brain circuits. A special area in the male brain's hypothalamus, the dorsal premammillary nucleus, or DPN, has been discovered, in rats, to contain circuitry for this instinctive one-upmanship. In humans, this one-upmanship and drive for status-seeking is found in men worldwide; it's not just a habit or a cultural tradition but more like a design feature of the male brain.

Neil's drive to maintain and gain status was occupying his every waking—and sleeping—moment. Danielle said she'd never seen Neil this sullen and angry. That's not surprising, for his testosterone was surging. And even if he wouldn't admit it, his brain knew that the confrontation would require more anger and aggression than he was accustomed to. Under normal circumstances, Neil preferred to be calm and relaxed, but he was willing to endure some nasty emotions if that's what it took to win the job he knew should be his.

The day of his first interview, Neil wasn't well rested, but he was determined to fake it. He put on a crisp white shirt and his red power-tie; he must look confident and in charge. When I saw him early that morning he looked sharp, and his jaw was firmly set. Testosterone clearly had activated his brain circuits and his manly facial muscles for dominance and aggression. Neil was in fight mode, and as far as his brain was concerned, this was war.

If we could peek back into Neil's brain in this atmosphere of unstable hierarchy, we'd see what was causing his emotional roller-coaster ride. When he thought his prospects for VP looked promising, we'd see his brain area for anticipating rewards activating, and he'd feel good. But when he thought George might get the promotion, we'd see his territoriality circuits in the DPN activating, and he'd feel haunted by the threat of losing face and forfeiting his place in the hierarchy.

The competition at work had become vicious, and Neil was obsessed with defending his territory. As he sat down in my office, he said, "The highlight this week was when my new boss, Ben, finally got fed up with George. He usually laughs at George's sarcastic comments. But yesterday George interrupted him in a meeting and Ben shot him a look that could kill. It was awesome." I was glad to see Neil in a more confident mood. As he left, he said, "My final interview for VP is next week. Wish me luck!"

It was several long weeks before Neil was finally offered the VP position, and when it happened Danielle and I both breathed a sigh of relief. But no one was more relieved than Neil. He could finally get some sleep. For Neil, the fight wasn't just about being the new VP; it was about beating the usurper, George, and defending his place in the hierarchy. By re-establishing a stable pecking order with himself at the top, Neil had achieved another milestone in manhood and set himself up for decades of continued career success.

SEVEN

The Mature Male Brain

JOHN, A fifty-eight-year-old business consultant, looked younger and more fit than when I'd seen him five years earlier. At that time, he'd been going through a difficult divorce, and the stressful side effects had been written all over him. Now he not only seemed more relaxed, but he exuded the self-assurance of a man who had finally come into his own—and knew it.

What was different about John at age fifty-eight? Nothing and everything. He had the same personality and brain circuits he had in his thirties. But now his highly responsive Maserati male brain—built for pursuit, competition, and aggression—was starting to run a different fuel mixture, more suited to a luxury sedan. And he was starting to enjoy his slightly slower pace. This difference is a normal part of the mature male brain, initiated by a shifting ratio of hormones. And as his hormones shifted, so would his reality.

John sat down in my office with a sigh and explained that

he'd been dating an interior designer named Kate for the past six months. He said, "Everything's going great with Kate, but my oldest daughter, Rachel, is really upset. I'm not sure if it's because Kate and I are getting more serious or because Kate's just six years older than Rachel." At this point, John started shifting in his chair and raking his hands through his salt-and-pepper hair. Frowning, he said, "Kate is young, but I've never met anyone like her before. I feel so close to her already."

John said that with Kate he even liked doing things that he used to avoid with women, like holding hands and cuddling. The tenderness he felt for her was new for him. Men in their fifties and sixties, like John, are beginning to make less testosterone and vasopressin, and researchers have shown that the ratio of estrogen to testosterone increases as men get older. Hormonally the mature male brain is becoming more like the mature female brain. Some scientists believe that with a different balance of fuels running a man's brain, he may become more responsive to his oxytocin, the cuddling and bonding hormone.

As a matter of fact, in studies in which researchers gave oxytocin to men, it improved their ability to empathize and enhanced their capacity to read subtle facial expressions. (Welcome to *our* world, guys!) Thus, because older men have lower levels of testosterone and vasopressin, estrogen and oxytocin may have a more dramatic effect on them. It certainly seemed that way for John, now that he'd met Kate.

If we could have watched John's brain as he sat across from Kate at their favorite restaurant, we'd have seen many circuits responding just as they had in his thirties. His visual cortex would be lighting up and registering how beautiful Kate was.

And when she looked at him admiringly, his RCZ—rostral cingulated zone, the area that registers others' opinions of us—would have activated, signaling that she looked up to him. When she toasted him for winning a competitive business contract, we'd see his reward circuits, the NAc and the VTA, pulsing with activity. But as he reached across the table to touch Kate's hand, he first carefully studied her face and gazed into her eyes, looking for clues about how she really felt.

As dinner continued, his visual cortex would activate again and again as his eyes returned to her full lips and slender neck. Soon we'd see his sexual brain circuits light up. And as he looked forward to the night ahead, we'd see his reward circuits flashing brightly with eager expectation.

But as the topic turned to his daughters and the future, we'd see his ACC, the anxiety and worrywart center, activate. He said, "Kate, you know how I feel about you, but I have to wonder if our age difference *is* too big. You know, you have your whole life ahead of you."

When Kate firmly reassured him, "I don't want a younger man. I want you, John," we'd see his ACC calm down. And when she said, "I've never had anybody listen to me and understand me like you do. Guys my own age can't do that," we'd see his reward center turning back on. This was music to his ears.

KINDER AND GENTLER

John's changing hormone levels were ushering in a kinder, gentler man. He still lost his temper when he got stuck behind "Sunday" drivers, but overall he was a little more patient and tolerant. His mature male brain was beginning to see the world more as it had when he was a boy, before the hormonal changes at puberty stimulated his circuits for anger and defense. And because he now had less testosterone, his oxytocin was having a bigger calming effect on his brain. This makes men a little less territorial, and they no longer feel compelled to fight so hard for their place in the pecking order. At this age, men even begin to risk showing more emotions without being so concerned about losing face. And they may also find themselves becoming more physically affectionate.

While cuddling had always seemed silly to John, now that his oxytocin was having a greater effect, he liked it. The skin-to-skin spooning that Kate liked to do before going to sleep now made him feel warm and content. John was no longer the unemotional robot that his ex-wife used to complain about. His new brain fuels were laying the groundwork for greater intimacy. And he said the sexual intimacy was the best he'd ever experienced. He found that with a younger woman, his sexual desire was reignited without using porn for the first time in twenty years. He was surprised by how much he liked pleasing Kate sexually; it was almost more important to him than his own pleasure. This was also new for him. Now he could slow down sexually, be a better listener, and be more affectionate. This hormonal turn of events can make men John's age become more like the ideal man that women say they want.

ESTABLISHED RANK

John's mature male brain was also changing his attitude at work. He had achieved the status of top dog in his industry years earlier, and now he could coast a little. His brain circuits for dominance and the drive to outdo other men were less intense as his testosterone production declined. He still fought for the Fortune 500 accounts, but victory at all costs was no longer worth it to him. During a man's midlife transition, he often isn't as motivated about fighting his way up the ladder. He knows his value. This development, which is usually attributed solely to psychological maturation, is also fostered by a new biological reality.

This also happens in other mature male animals when they reach this stage of life. Researchers have found that mature alpha-male silverback gorillas provide protection and leadership, maintain group stability, and mediate conflict. And they continue to offer companionship and protection to females long after their breeding years with them have subsided. Researchers found that the females in groups without a mature male silverback felt less protected and were less safe at night. So, instead of sleeping in their preferred, comfortable ground nest, these females had to sleep high up in trees for protection. Human females also can feel an attraction to the protection and security that a mature male offers. And for this reason, when it comes to choosing a mate, many younger women see the benefits of a well-established older man.

MEN STAY FERTILE FOR LIFE

When John brought Kate to my office to talk about smoothing things over with Rachel, it was easy to see why he was so enchanted by her. Kate was a lively five-six brunette, with a trim waistline, generous bustline, and pretty face. And it was clear by the way she looked at him with adoration that she was devoted to him. What John probably didn't know was that his mature brain found Kate's adoration almost as enticing as her breasts. He was being biologically bewitched to bond with her. And Kate was bonding with him, too. But for her female brain, bonding meant more than being lovers.

We were discussing the importance of giving Rachel time to come around when Kate said enthusiastically, "Of course we have to do what's best for John's daughters, just like we'll do what's best for our own kids someday."

My mind went blank for a moment. I had heard nothing from John about starting a new family, so feeling a bit tongue-tied, I turned to see his response.

The look on his face said it all. This was news to him, too.

Suddenly my office felt too small. All three of us knew this could be a deal breaker. Men who have already raised one family are often reluctant to do it all again. But some men do just that. We've all seen older men pushing baby strollers and wondered, *Is he the father or the grandfather?* The truth is that men can have a second, third, or fourth chance at fertility that women simply don't have.

In fact, the ability of older men to reproduce with younger women, called the "late-life male fertility factor," may be partly responsible for our species' long lifespan. Scientists at Stanford University found that this factor explains why humans live so long, even though, in theory, a female has outlived her evolutionary purpose by midlife when she is no longer fertile. Because men and women share most of the same gene pool, they both potentially benefit from any longevity genes in the other sex. Scientists argue that it's these late-life fertility genes that women share with men that explain women's longer lives.

At fifty-eight, John was certainly still fertile. But he and

Kate had plenty to think about, not least of which was working things out with Rachel—if they stayed together. This was a major decision for John, so I wasn't surprised when he told me a couple of weeks later that he was losing sleep over it. He said, "I'm crazy about her, but I don't want more kids. And Kate won't be happy without them. I've decided I have to let her go."

THE LONELY HEARTS CLUB

A few months after John told me that he and Kate split up, I got a call from his other daughter, Mandy. "Rachel and I are both worried about Dad," she said. "When he's not working, he mopes around the house. When I ask him what's wrong, he just gives me 'that look' that means the conversation's over."

John didn't want to go out by himself, and he hadn't met anyone new, so he was spending most of his evenings at home alone. Since he'd stopped having fun, his brain's reward centers were short-circuiting. John's hermit lifestyle wasn't giving his brain the social workout it needed, and his daughters could tell it had become a vicious circle as he became more cranky and withdrawn. Without Kate around, his social-approval circuits weren't being activated either. In brain-scan studies of older males, researchers have found that the brain's pleasure and reward areas, the VTA and the NAc, remain more active in men who are social.

The next week, when John came in to see me, he said, "My daughters think I'm depressed, but I don't think I am. I'm fine when I'm working. It's only when I go home to an empty house that I feel like crap. I know it sounds pathetic, but the only time I don't feel lousy is when I'm remembering the good times I had with Kate. But *then* I feel even worse because I don't have her anymore. You know I've never been one to mope around much."

Like John, many men who are lonely think it's a weakness, but it's actually a key survival mechanism. Mother Nature has purposely wired the state of loneliness into the human brain to cause pain so humans will avoid it. In primitive cultures, being isolated from your tribe could be a death sentence, because individuals could rarely survive on their own. And in today's modern world, researchers are finding that loneliness can still be deadly. People who are lonely die sooner than their same-age peers who are not lonely. They found that about one in five Americans experiences loneliness and that it can be as detrimental to your health, in the long run, as smoking.

When men live alone and become isolated—which they do more often than women—their daily routines can become repetitive habits that get deeply engraved into their brain circuits. Soon, if someone disrupts their routine, they get irritated because their brain's social-flexibility circuits are weakened from disuse. This is the story of grumpy old men.

It's also the story of grumpy old mice. Researchers have found that male mice need females around to keep certain other circuits active. It has long been known that females can influence a variety of responses in male physiology and psychology. Males of many species fare better with female companionship. And researchers found that older male mice that were living with females retained their reproductive abilities longer. But it isn't just the male gonads that can benefit from female companionship—it's their brains too. In humans, researchers found that specific brain circuits are not activated as much in people who are socially isolated. When brain areas aren't used enough, they atrophy. Isolation is bad for the brain. And although John wasn't a grumpy old man yet, loneliness was clearly bad for him.

I watched an array of emotions flash across John's face as he processed what I had just said about his brain, his gonads, and loneliness. And then he said, "Maybe having another baby or two wouldn't be so bad after all, if it means being with Kate the rest of my life."

What John didn't know was that having a baby with Kate was probably the best way for their brains and bodies to biologically pair-bond and stay together for the long run. Regardless of their age difference, Kate and John would be working *with* their pair-bonding brain circuits by starting a family. And soon, once they were back together, his biggest concern became staying healthy so he'd be around to raise and take care of his new family. Thus, I shared with him that, aside from inheriting good genes, a man's best chance for longevity is to sleep deeply, stay strong, avoid tobacco, and get married and stay married. For unknown reasons, married men live 1.7 years longer than single men. But as my patient Tom joked when he and his wife, Diane, were in marriage counseling, "Those extra years better be damned good!"

BIOLOGICAL CHANGE OF LIFE

Tom and Diane made a handsome couple: he with his trim physique, graying hair, and healthy complexion; she with her petite athletic figure, bouncing pageboy haircut, and sparkling eyes. They had initially come to see me when Diane's hormonal transition at menopause had decreased her sex drive. Her female brain was suddenly getting a lot less sex hormones, and she was experiencing a new biological reality that her husband couldn't relate to. Tom had always been a loving husband and father, but what he didn't know was that the combination of Diane's hormones and his lack of attention was making things worse with her libido. Diane had been angry because Tom's work always seemed to come before her. During the menopause, one wrong word or even just a look from Tom could send her slamming doors throughout the house and taking refuge in her greenhouse for a private sobfest. Her interest in sex was deader than a doornail.

On the other hand, Tom had long resented that Diane didn't appreciate how hard he worked to provide her and the kids with the lifestyle she wanted. Only when Tom was at his wit's end from being sexually rejected by her had he finally agreed to come to counseling. Working through their issues at that time had helped them to decide on some compromises and to renegotiate parts of their "marriage contract." Tom agreed to work less, and Diane agreed to be more attentive to his sexual needs. For some reason, many couples think they can't renegotiate the unwritten marriage contract—or "revisit the prenup," as I put it. To them I say, "Of course you can. Your life depends on it."

For Tom and Diane, the couples counseling and Diane's hormone-replacement therapy made all the difference. Consequently, I hadn't seen them in a few years. But now Diane was calling to say she thought Tom had better come in to see me, this time about *his* hormones.

Hormones in the brain and the penis are what make a man a man. They activate the sex-specific brain circuits required for male-typical thoughts and behaviors. When the male brain and body start making less of these hormones, he enters the so-called andropause, or male menopause. A century ago, andropause was relatively rare, because men didn't live long enough to experience it. Even in the late nineteenth and early twentieth centuries, the average age of death for men in the United States was forty-five. Nowadays, thank goodness, there's plenty of life after andropause. Men in the United States can expect to live decades after their hormones start to decline. But according to researchers, many men aren't happy unless this stage of life also includes sex. And for Tom, this was where the rubber met the road.

Over the phone, Diane explained to me that there were a few times lately when Tom couldn't get a firm enough erection to have intercourse. She said, "I personally don't think it's that big a deal. I mean he *is* sixty-four. The first two times it happened, he was upset about it, but he thought maybe he was just tired. But when it happened last night, he got up, got dressed, and stormed out to his woodshop. Do you think it could be hormones?"

Based on Diane's description, I agreed that Tom's testosterone could be low, but it could also be many other things. So I suggested that he come into the office to discuss what was going on and see how I might help.

Women can't imagine how changes in sexual function can affect a man's self-perception so much at this stage in life. He's the same man he's always been, and yet he's not. The body part that has brought him the most pleasure, the very symbol of his manhood, may no longer be reliable. As Tom sat down in my office, I could tell this was clearly taking a big toll on him. He said, in his usual get-down-to-business manner, "I'm sure Diane's filled you in. What do you think I should do?"

Tom was in good health, so I wanted to know what else might have changed since the trouble in the bedroom began. He said his last checkup and tests with his primary-care doctor were normal, even though he was feeling tired all the time and his thinking wasn't as sharp as it used to be. He complained that he'd lost strength, even though he exercised, and he noticed that his beard wasn't growing as fast as it used to. Otherwise, he said, he was "as good as ever." After hearing Tom's symptoms, I suspected that his testosterone level might be low and recommended that he get it tested.

The hormone tests confirmed that Tom was going through andropause. On average, men go through this hormonal transition between age fifty and sixty-five. During this stage, their testicles produce one third to a half the testosterone that they produced in their twenties. It was a good thing Tom came in, because this was a biological problem that was quickly becoming a marital problem.

As it turned out, Tom's testosterone level was substantially lower than average for a man his age. Researchers have found that when this hormone dramatically decreases, the brain and spinal cord don't get as much stimulation as they need to keep his sexual thoughts and organs working at full capacity. At all ages, the male brain, spinal nerves, and penis need testosterone to function.

Although Tom was discouraged by his penis's lack of reliability, he confided that on the positive side of the coin, he didn't mind being a little less sexually driven. He was glad that the "I need it now or I will die" urges that he used to have in his younger years had subsided. Now he could wait. But when the waiting was over, he wanted to be ready for action.

He said, "I was hoping that staying in shape would keep

things going down there. Now I'm not so sure. My brother's doctor started him on testosterone replacement a few months ago, and he swears by it. Does that stuff do any good?"

Although studies in the United States are few and far between, researchers in the Netherlands found that giving testosterone to men with abnormally low levels improved the men's physical and mental health. They found that it revived libido and penile function too. In addition, the men in the study had better muscle elasticity and bone density. They also reported improvements in their mood and their cognitive ability. And as a bonus, the study showed that increasing testosterone can kick-start abdominal weight loss.

I told Tom that research does show that exercising and staying sexually active can help men make more androgens, like dehydroepiandrosterone (DHEA) and testosterone. But for some men, that's still not enough. Being tired all the time and having foggy thinking aren't a natural part of getting older. When those symptoms are caused by abnormally low levels of androgens, some men get good results from androgen-replacement therapy. But its safety is still controversial, and it's not for everyone. For many men, there may be other answers.

In a study I conducted at UCSF in 1996, I compared the replacement of DHEA, an androgen similar to testosterone, with placebo in men over sixty. During this yearlong study, for an entire day once a month, the male subjects would come to UCSF's pleasant, spa-like clinical facility for testing. My friendly female nurses and psychologists would spend the day with the men while administering cognitive tests, collecting blood samples, and discussing details about their sex lives. Our results showed that by the end of the year, men taking the DHEA had improved their cognition, well-being, and sexual function by an impressive 40 percent. But the big surprise was that those taking the placebo had improved in all those same functions by 41 percent! I was forced to conclude that the therapeutic power of monthly social interaction with a group of caring, interested females was just as potent—or more so—to these andropausal men as the hormone DHEA.

A TIME FOR EVERY PURPOSE

When I told Tom about my study, he immediately understood that spending quality time with Diane might help him as much as hormone replacement. He said, "When Diane was going through menopause, the things you explained to us and encouraged us to do together made a big difference. But the hormones helped her too. How about if Diane and I come in to see you for some couples therapy again and I try the testosterone replacement too?" Tom wanted to keep his marriage and sex life going strong, so he was enthusiastic about doing whatever it took. And research confirms that men still rank sex as the major benefit of marriage, even into old age. So after discussing the risks and benefits with his primary-care doctor, Tom scheduled his first injection that very day.

When Tom and Diane came into my office together the next month, he explained that the testosterone-replacement therapy was helping but his "reliability" had slipped another notch. He said, "I don't want to be like this for the rest of my life. So I took my brother's advice and asked my doctor for Viagra."

Diane laughed and said, "I like the idea of Tom being happier with sex, but I'm not so sure I want him chasing me around the house again with that lustful look in his eye all the time."

Tom chuckled and said, "Okay, I'll only chase you around the house on weekends. The other days, I'll just chase you around the bedroom."

I could see Diane was happy that Tom was finding it easier to perform reliably again. And when I told her that drugs like Viagra, odd as it may seem, might even make him want to hold hands, cuddle, and hug more, her ears perked up. Researchers at the University of Wisconsin surprisingly discovered that drugs like Viagra can increase the release of oxytocin in the brains of rats by as much as threefold. Perhaps the famous blue pills will someday be used to promote deeper emotional intimacy and not just better erections.

Furthermore, as I reminded Tom and Diane, they could increase their oxytocin naturally by holding hands, gently stroking each other's skin and hair, massaging each other, and gazing into each other's eyes. Researchers showed that warm touching between partners improves satisfaction in relationships, and can actually make more of a health difference for men than women. In a "warm touch" study in Utah, researchers found that while both husbands and wives got a boost in their oxytocin levels and a reduction in stress chemicals, only the husbands got a beneficial reduction in blood pressure.

I also reminded Tom and Diane to practice the five-to-one rule: giving each other five compliments for every one critical remark. This was very important, because although Diane wanted Tom to be more affectionate, she admitted she had become more critical of him over the years. I said to Diane, "If you want Tom to hold hands with you and be more romantic, you have to be nice to him. And nice doesn't just mean sex."

Like Diane, women who have been married a long time know all their mate's foibles and flaws. The female brain tends to run negative scenarios to protect itself from disappointment and then place the blame on the male brain, like pinning the tail on the donkey. Constant criticism takes its toll on the brain. When a man's partner is critical of him, his brain goes on the defensive. His RCZ tells him that he isn't meeting her mark, and he begins to avoid contact. This creates a downward spiral that is a guaranteed dead end. It stops them both from getting the love and caring they crave. I knew using the five-to-one rule could make a positive difference for both Tom's and Diane's brains. And fortunately, in this case, the age of their brains would be working in their favor too.

According to scientists, older people don't let bad news or criticism get to them as much as younger people do. In a study comparing how the brains of seventy-year-olds versus twenty-seven-year-olds handle negative emotions, the older adults performed better. In particular, the researchers found that the brains of older adults had developed greater connectivity between the PFC, the area for regulation of emotions, and the amygdala, the area for driving emotional impulses. They concluded that the mature brain was not only better at controlling negative emotions but better at letting go of them. I said to Tom and Diane, "So, you see, it turns out that older brains are better at forgiving and forgetting." Tom quipped, "Maybe wisdom *does* come with age."

When I next saw Tom and Diane, they had just returned from a three-week European vacation. He gave me a thumbs-up, and she had a twinkle in her eye that I hadn't seen in years. It was nice to see them shifting comfortably into the next phase of their lives together.

THE GRANDFATHER BRAIN

The years after andropause are a major transition for the male brain. The fuels that are running a man's brain circuits change more toward oxytocin and estrogen and away from vasopressin and testosterone. Men are also winding down careers and looking for new and interesting projects to keep them busy and "in the game," or at least active on the sidelines. Up until maturity, many men can feel burdened by family commitments and even by their intimate relationships. But after andropause, they may, like Tom, have the time and temperament to appreciate family and friends more. Tom told me he felt a deep sense of satisfaction in having raised an accomplished daughter and that both he and Diane loved spending time with their grandson, Tommy.

To the amazement and delight of Diane and their daughter, Ali, Tom had recently morphed into a doting grandfather. Tom's mature brain had more patience with his grandson than he had ever had when Ali was growing up. The difference between Tom's daddy brain and his grandfather brain was working in everyone's favor. Much to even Tom's surprise, the love circuits in his mature brain were being hijacked by the adorable little Tommy, even more than they had been when Ali was born. The highlight of Tom's week had become cheering his five-year-old grandson on at his T-ball games. Soon, not even golf was as much fun as spending time with Tommy. He was amused by everything his grandson did and said. And he felt a deep pride in all of Tommy's developmental milestones and daily achievements.

One of my mentors, George Vaillant, found in his ongoing study of men who graduated from Harvard in the 1950s that in later years they changed their focus from activities that gave them a personal advantage toward activities that would give their community and the next generation an advantage. He calls this stage the fifth stage of individuation, or the stage of generativity. For many men like Tom, this position of a "wise elder" in their community includes the grandfather role, and a heightened interest in supporting the success and survival of the next generation.

Evolutionary anthropologists argue that grandfathers have been important for our species' survival. They found that in hunter-gatherer cultures, grandfathers can produce or procure more food than they consume, and therefore aid the flow of food from old to young. We now live in a world where the intelligence and knowledge of the grandfather—along with his financial assets, the modern equivalent of food—get passed on as his legacy to his kids, grandkids, and community. (In a typical example of the modern flow of resources from older generation to younger, Tom and Diane were helping Ali and her husband pay Tommy's private-school tuition.) But not all men find themselves particularly eager to take on the grandfather role. Researchers have shown that, at first, many men accept responsibility for grandchildren only out of a sense of obligation and love for their adult children. And research highlights how complex the nature of the ties between grandparents, their adult children, and grandchildren can be. For example, researchers have found that parents serve as gatekeepers and mediate the amount of contact and the quality of the relationship between grandparents and grandchildren. So a man's bond with his grandchildren depends on his relationship with his adult children. Tom was happy that his daughter facilitated his closeness with Tommy.

In many cultures, grandfathers also function to informally teach motor skills to grandchildren, for children learn best by imitation. And for Tom, not only did he love teaching his grandson how to throw and catch a baseball, he enjoyed teaching him about saving money and bought him his first piggy bank. He remembered how he got his own work ethic and the advice to save and invest his money from his immigrant grandfather, and he hoped to pass this on to Tommy. By the time his grandson was approaching his sixth birthday, Tom felt closer to him than he ever imagined was possible. He said, "If you'd have told me ten years ago that one of the highlights of my life would be having a grandson, I'd never have believed it."

Now that my husband, Sam, and I have entered this stage of life, with adult children and grandchildren of our own, we know exactly what Tom means. We find ourselves excited to interact with the grandkids as often as possible, chatting online by Skype and taking trips to see them. We feel more motivated than ever to make a difference in the lives of our family and the students and young faculty we mentor and teach. Sam and I have renewed our friendship and commitment as children have left home, come back home, married, and had grandchildren, and as a result, we are constantly finding a new balance. Of course, I can't see into our future, but the decades ahead beckon with hope, adventure, and passion for both me and Sam—the male brain I know best.

EPILOGUE

The Future of the Male Brain

IF I COULD impart one lesson to women that I learned in writing this book, it would be that understanding the biology of the male brain helps us relate better to the male reality. Much of the conflict that exists between men and women is fueled by unrealistic expectations that stem from failing to grasp each other's innate differences. To men, I would say I hope that shedding light on the male brain's tendencies and its physiological responses to hormones will clarify the basis for your natural urges and the way you think, feel, and communicate. My belief is that this information can provide men with a sense of relief at finally being understood.

Most people, men included, believe that the male brain's overriding goals are sex, status, and power—not necessarily in that order. And it *is* true that a tendency to seek these ends is built into the male brain's circuitry. But that is far from the whole story. Boys from the get-go learn differently than girls do and are interested in different things. Action, assertiveness, and rough play are biologically wired. We joke that men are run by their libidos, but the reality is that they are not slaves to their testosterone or sex drive. As we've seen, a man's sex drive may mature into a capacity for love and attachment that is at least as strong as a woman's. The stereotype of the stoic, unemotional male is again contradicted by research showing that the daddy brain and mature male brain are profoundly devoted and nurturing. And males don't start off less emotional. As infants, boys are actually more emotional than girls. But soon social pressures, child-rearing practices, and biology begin reshaping the male brain circuits. Teaching boys to suppress their feelings and facial expressions, along with testosterone's influence, becomes "successful" by the time they reach manhood. This is a result of both training and biology. The male brain's response to emotions in others soon takes its own path, producing fix-it-fast solutions designed to alleviate distress.

On a personal level, I believe that learning about the male brain can help men and women feel more intimacy, compassion, and appreciation for each other. Such understanding might be the most important factor in creating a genuine balance between the sexes. I hope that this book will foster such understanding and help create more humane and civil societies wherever it is read.

APPENDIX

The Male Brain and Sexual Orientation

Are men gay because their brains are different? Studies have been conducted for two decades in order to answer this question. Some of them have found evidence of anatomical or functional differences between gay and straight brains. Others have established that genes play a part in determining gender orientation, which implies the existence of brain differences.

One of the early studies, by Dick Swaab, found that a part of the hypothalamus called the suprachiasmatic nucleus (SCN) is twice as large in gay males as in straight males. This difference was later shown to be caused by a difference in the way testosterone reacts with the developing brain. Other researchers showed that the anterior commissure—a bundle of superfast cables that connects the brain's two hemispheres—is larger in gay males than in straight males. This structure, which is also larger in women than in men, is believed to be involved in sex differences related to cognitive abilities and language, and fits with the finding that gay males, like females, have better verbal abilities than straight males.

Recently Ivanka Savic and other Swedish researchers reported that an anatomical asymmetry in the size of the two brain hemispheres that is characteristic of straight male brains is not observed in gay male brains. Instead, their magnetic resonance imaging studies showed that in this respect gay male brains were more like female brains. With PET scans, the researchers also found that the connectivity of the amygdala of the gay male brain is more like that of the straight female brain than of the straight male brain. These studies suggest that there are differences between gay and straight male brain areas that are not directly involved in sexual attraction.

Savic has also reported a different pattern of activation in gay male brains in response to a pheromone that is excreted in male perspiration. She found that the hypothalamus in gay male brains is stimulated by the scent of male sweat, but in straight male brains it is not. This suggests that a difference in the brain's hypothalamic circuits for response to pheromones may attract gay males to the scent produced by the sweat glands of men and that this plays a role in their sexual orientation. Other studies have found anatomical differences in the structure of the hypothalamus in gay and straight men.

There is also evidence of differences between gay and straight men in performance on certain spatial tasks. It has been consistently shown that straight men outperform straight women on tasks requiring navigation. Recent studies have shown that gay men perform more like straight women on such tasks.

Brain scans have also been used to measure activity changes in the gay and straight male brain when pictures of men and women were shown. Viewing a female face produced a strong reaction in the thalamus and medial prefrontal cortex of straight men but not of gay men. In contrast, the gay male brains reacted more strongly to the face of a man.

Genetic studies have also contributed some evidence of innate differences between gay and straight males. In a recent study, Dr. Niklas Långström estimated the part that genes play in gay male behavior by studying sexual orientation in adult male twin pairs. He found that identical twin pairs, who have all the same genes, are more likely to share sexual orientation than fraternal twin pairs, who share only half of their genes. Based on this comparison, he concluded that about 35 percent of sexual orientation is attributable to genetic influences, whereas the rest is due to as yet unidentified factors.

So far, none of the specific genes that influence sexual orientation have been identified, and researchers believe that the combined actions of many genetic and environmental factors will be involved. Furthermore, research on the brain circuits and on hormonal effects related to sexual orientation in humans is only just beginning. Nevertheless, the evidence we already have does show that some human brain differences are related not only to gendered behavior, but also to sexual orientation.

NOTES

INTRODUCTION: WHAT MAKES A MAN

1 *What Makes a Man:* Certain brain areas and functions are built differently in the male and female brains and have evolved over time to produce the most successful versions of men and women. For example, the brain circuits that alert us to danger (the amygdala) and help us remember it (the hippocampus) are the sources of sex and individual differences in emotional memory. In Hamann 2005, the authors found sex differences in the amygdala response during emotion-related activities, such as formation of emotional memory and sexual behaviors. And for more on the evolution of male and female brain circuits, see Lindenfors 2007 and Dunbar 2007: they say ". . . there are striking differences between the two sexes in the social mechanisms and brain units involved. Female sociality (which is more affiliative) is related most closely to neocortex volume, but male sociality (which is more competitive and combative) is more closely related to subcortical units (notably those associated with emotional responses). Thus different brain units have responded to different selection pressures." For more on cellular and genetic differences in the male and female brain, see Reinius 2008 and Arnold 2009b.

2 *and the fix-it-fast emotional brain:* Coates 2009 found that testosterone sets the male brain up for faster visuomotor scanning, faster physical reflexes, and more risky behavior.

2 *Male and female brains:* Penaloza 2009. The authors say, "Sex of the cell

dictates its response." Malorni 2007 even found reduction-oxidation differences between male and female cells.

2 *for later amplification by hormones:* For more on sex hormones, genes, and the brain, see Arnold 2009c and Neufang 2009.

3 *differences between women and men:* For good reviews on sex differences in the brain, see Becker 2008b, McCarthy 2009, and Proverbio 2009.

6 *role in shaping and reshaping our brains:* It is important to note that biological predispositions can be shifted over time through experience and that existing dispositions can be mitigated or even overridden by situational demands in both men and women. For both males and females, upbringing, experience, and the environment can make long-lasting biological and behavioral impacts via epigenetic changes to our DNA. See Merzenich 1983 for early work on brain reorganization in response to changed circumstances. For more on changes in brain architecture with experience, see Kozorovitskiy 2005, and on environmental experience and epigenetic effects, see Meaney 2005, McCarthy 2009, and Murray 2009.

ONE: THE BOY BRAIN

10 *wired into the male brain:* Coates 2009 found prenatal androgens promote increased risky behavior, movement, and physical reflexes. For more on the male brain, see Arnold 2009c, Van Nas 2009, Chura, 2010, Wu 2009, Field 2008 and 1997, Baron-Cohen 2003 and 2009, Pfaff 2002, Holden 2004, Eme 2007, and Becker 2008b; see also De Vries 2008 and McCarthy 2009a. In humans and most other mammals, a gene on the Y chromosome, the SRY gene, confers maleness. Studies suggest that the SRY gene directly affects the biochemical properties of the dopaminergic neurons of the nigrostriatal system and the specific motor behaviors they control. This means that a direct male-specific effect on the brain is caused by a gene encoded only in the male genome.

10 *their genes and sex hormones:* Arnold 2009a.

10 *on and off different genes:* Arnold 2004 and Wu 2009.

10 *play at fighting off enemies:* Auyeung 2009b says, ". . . our data are the first documentation that androgen exposure prenatally relates to sexually differentiated play behavior in boys and in girls." For more on genes and hormones in boys and girls, see Wu 2009 and Berenbaum 2008.

10 *seemed to find them fascinating:* Connellan 2000. For more on gender differences in newborns, see Ashwin 2009, Baron-Cohen 2009, Auyeung 2009, and Gilmore 2007.

11 *stereotypical male behaviors like roughhousing:* Wang P. 2009.

11 *in different activities than girls:* Maccoby 1998 and Byrd-Craven 2007.

11 *they begin in the brain:* Wu 2009 and Reinius 2008.

11 *upset, they're harder to soothe:* Weinberg 1999.

12 *when she was a baby:* Leeb 2004.

12 *without as much mutual gazing:* Leeb 2004. For more on maternal attachment and bonding, see Young 2008, Baron-Cohen 2003, Carter 1998, Nichools 1996, and Bowlby 1980.

12 *objects from the get-go:* Connellan 2000. For more on sex differences, see Hampson 2008 and Quinn 2008.

12 *"contact much more than girls":* By six months old, boys gaze-avert more frequently than girls. Whether they are straining to look at something else that caught their eyes or away from the face is not entirely known. For further reading, see Byrd-Craven 2007, Knickmeyer 2006, Bayliss 2005, and Hittelman 1979. For more on autism and the male brain, see Baron-Cohen 2009, who says that his results suggest that prenatal exposure to high levels of testosterone influences some autistic traits and thus hormonal factors may be involved in vulnerability to autism. Ashwin 2006 says that, for still unknown reasons, Asperger's and autism affect four to ten times more boys than girls. Autism and Asperger's syndrome (AS) are genetic neurodevelopmental conditions characterized by social deficits, abnormal face processing, and amygdala dysfunction that are more common in boys.

12 *"airplanes and other moving objects":* Moore 2008. For an overview of moving objects and movement in brain circuitry in boys and girls, see Hampson 2008 and Field 2008.

12 *weeks after he was conceived:* See Field 2008 for more on the development of sex differences in brain circuits for movement.

12 *circuits that control male behaviors:* For a review of the formation of male brain circuits, see Wu 2009, Gagnidze 2009, Becker 2008b, Eme 2007, Breedlove 1983, and Archer 2006. Note: a failure of masculinization is called *pseudohermaphroditism.* See Tsunematsu 2008 on vasopressin increasing locomotion in males.

12 *make others wither and die:* For more on hormones causing death or acting as growth factors, see Wu 2009 and Kimura 2008. For more on sex differences and brain development, see Penaloza 2009, Swaab 1985 and 2009, Ehrlich 2006, and Zuloaga 2008.

13 defeminized *David's brain and body:* Wang P. 2009. Wang and colleagues found that the foundations for the observed MIS-dependent increased movement and exploratory behavior in males is laid down in the

brain circuits during fetal life. Wang P. 2005 found that in males, Müllerian inhibiting substance is secreted by the fetal testes and causes the death of the Müllerian ducts in order to prevent the growth and development of female reproductive organs in male bodies. Note: MIS is also called "anti-Müllerian substance."

13 *off the female reproductive organs:* Normal development of the male brain involves two distinct processes, masculinization and defeminization. They occur during critical periods of brain sexual differentiation. Masculinization allows the expression of male sex behavior in adulthood, and defeminization eliminates or suppresses the expression of female sex behavior in adulthood. Once inside the fetal brain, much of the testosterone is actually converted by the enzyme aromatase into estrogen. Ironically, then, it is the estrogen that helps masculinize and defeminize the male brain, working in concert with MIS. For more on sex and the brain, see Wu 2009, Wang P. 2009, and Becker 2008b.

13 *spatial skills, and rough play:* Wang P. 2009 found that male mice lacking MIS exhibit feminization of their spinal motor neurons and of their exploratory play behavior. They hypothesize that, along with testosterone, MIS may be a regulator of the sex-linked behavioral biases in the nervous system and brain toward more movement, pursuit, rough play, and exploration in males.

13 *not develop male-typical exploratory behavior:* For further reading, see Wang P. 2009.

13 *effects of testosterone or MIS:* Fynn-Thompson 2003. MIS is *not* present in the female embryo but is induced in females only after birth. For a review of the association between testosterone and aggression that occurs in boys, but not girls, during childhood, see Becker 2008b, Eme 2009, and Archer 2006.

13 *to appreciate her son's maleness:* Diamond 2006 found that the boy's sense of masculinity early on is shaped by factors like the mother's recognition and affirmation of her son's maleness, the role of the involved or uninvolved father, and the nature of the parental relationship.

14 *testosterone as in an adult man:* For more on the infantile, or perinatal, masculinization of adult sex behavior, see Wu 2009, Wudy 1999, and Wright 2008.

14 *exploratory behavior and rough play:* Wang P. 2009 found that boys older than one year have very low levels of testosterone, but they continue to have high MIS until the onset of puberty.

14 *when she's angry or afraid:* Peltola 2009 found that an enhanced sen-

sitivity to facial signals of threat emerges between five and seven months of age. This may reflect functional development of the brain mechanisms involved in processing of emotionally significant stimuli. Grossman 2007 found that seven-month-olds integrate emotional information across modalities and recognize common emotions in the face and voice. For more on brain processing of words, faces, and emotions, see Schacht 2009. For more on infant gazing at objects and adult emotional expressions, see Hoehl 2008.

14 *and can easily ignore them:* Rosen 1992. They found that by the age of twelve months, the intensity of the mothers' fear signals increased and were more intense toward boys, but boys ignored them more. For overview, see Maccoby 1998, Mumme 1996, and Becker 2008b.

14 *For girls, the opposite happens:* Researchers find that by one year of age, girls are more socially oriented toward their mothers than are boys; see Wasserman 1985, Maccoby 1998, and Byrd-Craven 2007. They are more wary than boys, according to Jacklin 1983 and Gunnar 1984. Zahn-Waxler 1992 found that infant boys cry less than girls in response to hearing the distressed cries of another baby. By twelve to twenty months, boys are less interested in seeking out and inquiring about the distress of someone who is hurt; for overview see Byrd-Craven 2007 and Leppanen 2001. Even as children get older, studies show that what is important to girls and boys, and meaningful to pay attention to, is different in terms of nonverbal emotional communications. For overview of sex-difference brain research, see Becker 2008b.

15 *of warning on Jessica's face:* For boys ignoring their mothers more, see Rosen 1992 and Maccoby 1998.

15 *in the room with them:* Rosen 1992 in "An experimental investigation of infant social referencing: Mothers' messages and gender differences." For more on gender differences in brain and behavior, see Maccoby 1998, Byrd-Craven 2007, Eme 2007, and Becker 2008b.

15 *at the risk of punishment:* Cialdini 1998a.

15 *sons as to their daughters:* Maccoby 1998.

15 *take risks and break rules:* Maccoby 1998. For more on sex differences in play behavior, also see Minton 1971 and Berenbaum 2008.

16 *many mothers wince, including me:* A mother's reaction to a boy's genitals may have a bigger effect than is commonly acknowledged, even in other mammals. Wallen 2009 found that in primates, maternal increased responsiveness to boys is best accounted for by the mother's reactions to the penis of her offspring.

17 *are "girl colors" like pink:* Feiring 1987 and Fagot 1985. For overview of girls' and boys' toys, see Pasterski 2005 and Hassett 2008.

18 *destroying, and seeking new thrills:* Maccoby 1998. For more on the brain and boys' thrills, see Byrd-Craven 2007, Manson 2008, and Becker 2009.

18 *more interested in cooperative games:* Berenbaum 2008. McClure 2000 found that during the late preschool and early school years, girls tend increasingly to play in small, intimate groups, whereas boys more typically form large, hierarchically organized groups focused on competition. For overview of children's gender cultures, see Sheldon 1996 and Maccoby 1998. Charlesworth 1987 found that girls got greater access to the toys by verbal bargaining rather than through physical pushing like the boys.

18 *while girls spend only 35 percent:* Lever 1976.

18 *twenty times more often than boys:* Maccoby 1998.

18 *to use as toy guns:* Maccoby 1998.

18 *and used beans as bullets:* Sheldon 1996 and Maccoby 1998.

19 *is being called a girl:* Blaise 2005 found that in early childhood classrooms, being called a "girl" is considered by boys to be one of the most shameful, polluting, and degrading insults of all.

19 *like girls' games and toys:* Feiring 1987 found that boys reject other boys who like girls' games and toys.

19 *and must therefore be avoided:* Feiring 1987 says that even in the years before exclusionary same-sex play, groups evolved in several studies of children 24 to 36 months old, where boys were able to identify their own sex's toys and activities better than girls did. For more on self-imposed sex segregation, see Pasterski 2005 and Maccoby 1998.

19 *doll and the wheeled toy:* Hassett 2008.

19 *toys more than other girls:* Servin 2003.

20 *muscle groups when they play:* Hassett 2008. Eaton 1986 says that boys' toys reflect their preference for using big muscle groups when they play, and sports reflect boys' preference for gross motor behavior and propulsion—of themselves and objects—from infancy onward. For overview of gendered play, see Berenbaum 2008.

20 *like car and plane crashes:* Iijima 2001 found that drawings by boys and girls showed clear sex differences. Girls tended to draw human figures, mainly girls and women, as well as flowers and butterflies. Girls used bright colors, such as red, orange, and yellow, and their subjects

tend to be peaceful and arranged in a row on the ground. Boys, however, preferred to draw more technical objects, weapons, and fighting, and means of transport, such as cars, trains, and airplanes, in bird's-eye-view compositions and in dark, cool colors, such as blue. See also Tuman 1999 on children's gendered art.

20 *is to determine social ranking:* Archer 2006.

21 *seen the humor in it:* DiPietro 1981.

21 *break his sense of self:* Eaton 1986.

22 *"have their kind of fun":* Maccoby 1998.

22 *form of a dopamine rush:* Becker 2008b says that sex differences in the regulation of dopamine, DA, in the brain's ascending mesolimbic projections may underlie sex differences in motivation and result in the expression of differences in motivated behaviors between boys and girls, men and women.

22 *and children self-impose sex segregation:* Maccoby 1998.

22 *mock-fight frequently; girls do not:* Maccoby 1998. Also see Eme 2007, Flanders 2009, and Becker 2008a on sex differences in motivation.

22 *if they placed at all:* Grant 1985.

23 *the girls did not show:* Feiring 1987.

23 *establish physical and social dominance:* Eme 2007 and Becker 2008b.

23 *"most important thing to be good at":* Boulton 1996.

24 *of the six-month study:* Edelman 1973 and Weisfeld 1987.

24 *levels than did the other boys:* Weisfeld 1987.

24 *the hierarchy at age fifteen:* Weisfeld 1987.

25 *of learning boys do differently:* See Benenson 2003 and 2009 for more on boys' choosing to ally with stronger peers, and for more on girls' need to maintain conflict-free friendships.

25 *Wii had become their favorite toy:* Hoeft 2008 found that these gender differences may help explain why males are more attracted to and more likely to become hooked on video games than females. For more on computer game playing and improving spatial skills, see De Lisi 2002, Feng 2007, Ginn 2005, Olson 2007, Heil 2008, and Wolbers 2006.

25 *brain linked to dopamine production:* Hoeft 2008 found that males showed greater activation compared to females in the mesocorticolimbic system—dopamine brain areas—which may account for gender differences in reward prediction, learning reward values, and cognitive state during com-

puter video games. For more on the male brain and dopamine, see Lavranos 2006 and Becker 2008b.

25 *even if he isn't moving:* Grafton 1997.

26 *control his own jumping muscles:* Orzhekhovskaya 2005 found males have more activity in neurons in the motor areas of the brain. See Cherney 2008 on "Mom let me play more video games."

26 *girls do in this way:* For more on the gender difference in movement biology, see Field 2008.

26 *and express themselves as well:* Ehrlich 2006 found that the boys frequently gestured about movement even when they did not talk about it; they found that gesture (but not talk) was associated with successful performance on spatial transformation tasks. Boys used their hands and body to convey an understanding of the spatial transformation task not found in their spoken explanations. And for more on girls catching up to boys on visuospatial tasks by practicing video games, see Feng 2007, who found that playing an action video game can virtually eliminate the gender difference in spatial attention and simultaneously decrease the gender disparity in mental rotation ability. For more on sex differences in movement and the brain, see Field 2008, Hampson 2008, Spence 2009, and Becker 2009.

26 *the meaning of that word:* For more on embodied cognition, see Siakaluk 2008, Thomas 2009, and DeCaro 2009. See Ullman 2008 for more on sex differences in the neurocognition of language.

26 *their advantage at spatial manipulation:* Keller 2009 found evidence for gender differences in the functional and structural organization of the right-hemisphere brain areas involved in mathematical cognition. Spence 2009 found that women can catch up to men on spatial skills by practicing single-shooter video games.

26 *object in their mind's eyes:* Hahn 2009. Also see Koscik 2009, Hampson 2008, Hugdahl 2006, and Clements-Stephens 2009 for more on mental rotation.

27 *differences between boys and girls:* Scientists agree that there are genuine between-sex differences in cerebral activation patterns during mental rotation activities even when performances are similar and that the sexes use different strategies in solving mental rotation tasks. Clearly males and females can both generate male- and female-typical behaviors, but their brains use different strategies to do so. For structural differences in the brain related to visuospatial abilities and language, see Hanggi 2009, Shaywitz 1995, Jordan 2002, Piefke 2005, Neuhaus 2009, and Hampson 2008.

27 *them to grasp its three-dimensionality:* Scientists believe that sex effects

reflect a difference in strategy, with females mentally rotating the polygons in an analytic, piecemeal fashion and males using a holistic mode of mental rotation. See Yu 2009, Heil 2008, Schoning 2009, and Hooven 2004 on testosterone and mental rotation.

27 *explanation without using any words:* Ehrlich 2006 found that before training, the response for the boys was to produce answers in movement and gesture on all eight problems; however, the response for the girls was to produce answers in movement and gesture on zero of the eight problems. After "gesture-training," the girls did much better. For further reading on gesturing and math teaching, see Goldin-Meadow 2009. For more on movement and math, see Broaders 2007, Terlecki 2008, Thomas 2009, Lorey 2009, Thakkar 2009, and Munzert 2009.

28 *without intervention, use them differently:* See Hampson 2008 for more on sex differences in visuospatial cognition and learning. For more studies that found sex differences in brain circuits used for mental rotation, see Nuttall 2005, Casey 2001, Jordan 2002, Peters 2006, Quaiser-Pohl 2002, and Parsons 2004.

28 *amounts of the pheromone androstenedione:* Hummel 2005. And see Larsen 2003 on the dramatic hormonal changes that begin before the body changes of puberty.

TWO: THE TEEN BOY BRAIN

31 *different from the preadolescent brain:* Yurgelun-Todd 2007 found emotional and cognitive changes during adolescence. Brain regions that underlie attention, reward evaluation, emotional discrimination, inhibition of impulses, and goal-directed behavior undergo architectural remodeling throughout puberty and into early adulthood.

31 *little androgen switches:* Swaab 2009 found that certain areas of the male brain have more receptors for testosterone, i.e., androgen receptors, AR, than the same areas of the female brain at birth but await its maximum behavioral activation at puberty with the surge in testosterone production from the testicles. Kauffman 2010 found that the onset of puberty in males and females is controlled differently. Testosterone levels rise at puberty, and these levels support mating and aggressive behavior. And not until males are required to care for offspring, or until old age, does their testosterone level decrease. For more on testosterone and the male brain, see Matsuda 2008, Wu 2009, Sato 2008, Neufang 2009, Becker 2008b, Ciofi 2007, Zuloaga 2008, Shah 2004, and Schulz 2006 and 2009.

32 *many other boys his age:* Christakou 2009 found in teens a recruitment of age-correlated prefrontal, PFC, activation in females, and of age-correlated

parietal activation in males, during tasks of cognitive control. Perrin 2009 and Giedd 1996 and 2006 found that brain development in teens differs between boys and girls, reaching its peak in girls one to two years earlier than in boys.

32 *soared twentyfold:* Larsen 2003. For more on testosterone, androgens, and adrenarche, see Nakamura 2009 and Peper 2009a.

32 *that emerge from his brain:* Halpern 1998 found that when teen boys go through puberty, there is a big change in sexual and aggressive thoughts. Scientists agree that testosterone is the main driver of sex differences in aggression. Aggression and testosterone are reviewed in Archer 2006 and Terburg 2009.

32 *lengthen and thicken his penis:* Larsen 2003. During puberty the non-erect penis doubles in length. In boys, the fastest growth spurt in height occurs three years after puberty onset.

32 *as those in girls' brains:* Swaab 1985 and 2009.

32 *the forefront of his mind:* Halpern 1998 found that higher levels of testosterone were associated with first sexual intercourse. As puberty progresses, testosterone acts more effectively in the teen boy brain's amygdala, hypothalamus, and spinal cord, stimulating mating behavior and copulation. Reviewed in Archer 2006, King 2008, and Becker 2008a.

33 *if they're turning into "pervs":* Testosterone not only saves cells from being killed off in the male spinal cord and brain's visual cortex; it primes the visual cortex to focus on sexually attractive females. Some men recall when their brain's visual perception changed at puberty and, almost overnight, all it took was the *hint* of a female shape to snap their heads around. In females, two out of three of these special sex cells in the spinal cord die due to lack of testosterone. For more on cell death in the spinal cord, see Nunez 2000 and Breedlove 1983. In gay male teens the brain begins responding to visual cues from same-sex faces, body parts, and pheromones of other males. Narring 2003 found that 2.9 percent of boys by age 16–20 report same-sex attraction.

33 *girls, which runs on autopilot:* During puberty, increases in dopamine in the sex center and copulation center increase sexual motivation and stimulate visual imagery. According to Witelson 1991, this brain area is 2.0 to 2.5 times larger in the human male than in the female. Becker 2008b found that increased dopamine in this brain area is linked to increased sexual motivation. For more on sexual motivation in the male brain, see Yeh 2009, Halpern 1998, Eme 2007, and Balthazart 2007.

33 *always on in the background:* Becker 2008a found that in the male brain after puberty, sexual "motivation and sperm production are switched to the 'on' position all the time."

33 *a companion hormone called vasopressin:* For more on vasopressin regulation by testosterone, see Pak 2009. Devries 2008 found that the vasopressin innervation of the brain shows perhaps the most consistent neural sex difference and that males have more vasopressin (VP) neurons and denser projections from these areas than do females, and that VP helps defeminize sex behavior in males. Researchers have found that sex differences in VP also match sex differences in social behaviors, for example, aggressive behavior. For reviews of hormones, sex, and behavior, see Becker 2008b, Gleason 2009, Forger 2009, and Pfaff 2002. For more on the opposite effects of vasopressin and oxytocin, see Viviani 2008.

34 *to his status or turf:* Testosterone surges at puberty, so it ensures that when needed under the stress of male-male competition for sexually receptive females or procuring the resources necessary to attract such females, young men will be ready. And at the same time these surges of testosterone will *lower the sensitivity to punishment* and increase the sensitivity to reward. For overview, see Archer 2006. For more on testosterone and behavior, see Dabbs 1996, Van Honk 2004, Handa 2008, Becker 2008a, and Evuarherhe 2009.

34 *living in status-conscious hierarchical groups:* Behrens 2009 explains how the brain develops networks for living in status-conscious groups and for hierarchical social behavior. They found that two distinct networks of brain regions are particularly active. The first area is for learning about reward and reinforcement. The second is a network that is active when a person must make estimates of another person's hostile versus friendly intentions.

34 *from the bottom as possible:* For more on testosterone, the need for dominance, and why some people strive for high status, whereas others actively avoid it, see Josephs 2006.

34 *that get them into trouble:* For more on risky decision making and good judgment in the brain, see Weber 2008. For more on men, risky financial behavior, and testosterone, see Dreber 2009 and Coates 2008.

34 *began when he entered puberty:* Giedd 1996 and Lenroot 2007 found that the total cerebral volume peaks at age 10.5 in females and 14.5 in males. For more on brain development in puberty, see Berns 2009, Herve 2009, and van Duijvenvoorde 2008.

34 *"late teens or early twenties":* For more on teen brain development, see Cameron 2005, Luna 2004b, Tiemeier 2010, Giedd 1996 and 2006, and Schweinsburg 2005.

34 *to focus on his studies:* For more on adolescent mental development, see Yurgelun-Todd 2007 and Ochsner 2004.

35 *his sex and aggression circuits:* Trainor 2004 found that a surge in testosterone following an aggressive encounter caused males to behave more aggressively in another encounter the following day. For more on testosterone and vasopressin, see Young 2009a, Neumann 2008b, Raggenbass 2008, Kajantie 2006, Schulz 2006a, Thompson 2006, and Keverne 2004. For reviews of neuroscience, sex, psychology, and testosterone, see Becker 2008a, Eme 2007, and Archer 2006.

35 *hormone, cortisol, would start climbing:* Williamson 2008.

35 *center—the amygdala—would activate:* The major role of the amygdala is in alerting the brain to danger, thus triggering fear and anxiety. Debiec 2005 found that in the amygdala, vasopressin (whose manufacture is stimulated by testosterone) and oxytocin (whose manufacture is stimulated by estrogen) work in opposite directions. For more on sex hormones and behavior, see Huber 2005, Pittman 2005, Donaldson 2008, Herry 2008, Tsunematsu 2008, Viviani 2008, and Bolshakov 2009.

35 *homework just doesn't do it:* Williamson 2008 found that cortisol, the stress hormone, starts to have less and less activating effect on the male brain as testosterone and dopamine rise. Thus it takes more and more excitement to get the brain's attention. For more on the down-regulation effects at puberty on the male brain by dopamine, see Becker 2008a.

35 *in tenth or eleventh grade:* National Center for Education Statistics. Also reviewed in Tyre 2008.

36 *high-school dropouts are boys:* National Center for Education Statistics. For more, see Dropout rates in the United States, 2004. National Council on Education 2009.

36 *of the Ds and Fs:* National Center for Education Statistics, National Council on Education 2009.

36 *eleven or twelve years old:* Roenneberg 2004.

37 *brains require at least ten:* Hagenauer 2009 found that this sleep deprivation is due to pubertal changes in the homeostatic and circadian regulation of sleep, promoting a delayed sleep phase for later bedtimes. For more on sleep disorders in teens, see Crowley 2007.

37 *to be excited about anything:* Becker 2008a: The baseline, or set-point, in boys changes, becoming less reactive. For more on emotional and cognitive changes during adolescence, see Yurgelun-Todd 2007.

37 *area in adults and children:* McClure 2004. And for more on pleasure, reward, and risk in the brain, see Bornovalova 2009.

37 *feel normal levels of stimulation:* Becker 2008a found that the mascu-

linizing and feminizing of the dopamine motivation system increases in the brain at puberty.

37 *much as children's or adults':* McClure 2004. For reviews, see Becker 2008b, Steinberg 2004a, Teicher 2000, Keating 2004, and Paus 2009.

38 *how different her outlook would be:* Response to threat and protection of territory change at puberty. Archer 2006 found that preliminary evidence from fMRI studies suggests a relationship between testosterone and amygdala responsiveness to anger in faces, perhaps leading to more aggression in males.

38 *perceives other people's facial expressions:* McClure 2004.

38 *by modifying our brain's perceptions:* Increasing hormone levels at puberty prime the brain for new behaviors. Even in childhood, hormones were acting to prime behavior, just at lower levels. For more on testosterone priming boys' behavior, see Archer 2006, Finkelstein 1997, De Vries 1998, Van Honk 2004, and Dabbs 1996.

38 *teen boy's sense of reality:* Thompson 2004. For more on vasopressin neurons whose projections extend deep into the male brain, see Caldwell 2008.

38 *way teen girls perceive reality:* See Carter 2009 for overview of oxytocin and vasopressin and social behavior in males and females. For more on the underlying neuroscience behind gender differences, see Becker 2008b.

38 *for aggressive and territorial behaviors:* Craig I. 2009, O'Connor 2004, and Archer 2009 and 2006. See Becker 2008a for perception and motivation changes in the male and female brains.

39 *squirt of vasopressin nasal spray:* Thompson 2006 showed that giving males extra vasopressin stimulates angry and competitive facial expressions in response to seeing the faces of unfamiliar males. For more on changes in male brain responses to faces from teen years to middle age, see Deeley 2008.

39 *potential threats when hormonally primed:* Motta 2009, Becker 2008a, and Gobrogge 2007.

39 *more territorial aggression and mate protection:* Gobrogge 2007 found that males who were pair-bonded to a female for two weeks displayed intense levels of aggression toward strangers. They hypothesize that dopamine and vasopressin in the hypothalamus may be involved in the regulation of enduring aggression associated with pair bonding in males.

39 *of facial posturing and bluffing:* See Sonnby-Borgstrom 2008 for more on facial posturing and gender differences in facial imitation, emotional contagion, and regulation of facial expression.

40 *is used to maintain power:* Archer 2006 found that in societies where the dominant male's display of anger enforces order, angry faces serve an important purpose. The battles for dominance in the primate kingdom, where the alpha male stares down his opponent and the loser male averts his eyes, are observed in humans, too—so much so that high testosterone is associated with staring endurance, or face-to-face dominance. Men with less testosterone have been shown to exhibit more submissive behavior by breaking eye contact, looking down and away.

40 *highest testosterone, according to research:* Archer 2006. Rowe 2004, who analyzed testosterone in boys aged 9–15, found that the dramatically increasing testosterone in pubescent boys contributed to social dominance and leadership status.

40 *reacted more aggressively to threats:* Olweus 1988.

40 *being more irritable and impatient:* Olweus 1988 found that high levels of testosterone made boys more impatient and irritable, which in turn increased their propensity to engage in aggressive-destructive behavior.

40 *the brain circuits for aggression:* Wirth 2007. For more on angry faces and testosterone, see Van Honk 2005 and Delville 1996.

40 *than he did before adolescence:* Rymarczyk 2007.

40 *between girls' and boys' brains:* Rymarczyk 2007 found a sex difference in the brain's processing of tone of voice. For more on sex differences in brain chemistry and the sex-determining gene located on the Y chromosome, see Wu 2009 and Paus 2009.

41 *they processed the sound of music:* Ruytjens 2007 found the male brain screened out white noise better than the female brain. For more on gender differences in auditory processing, see Voyer 2001 and Ikezawa 2008.

41 *than the female brain does:* Ruytjens 2007. For more on fetal brain development and the effects of testosterone on hearing, see Beech 2006 and Cohen-Bendahan 2004.

41 *musical banter was practically impossible:* Schirmer 2002 studied sex differences in neural processing of emotional words, and found the tone and meaning of emotional words were processed faster in females than in males.

41 *the hot substitute teacher's measurements:* Guiller 2007 found gender differences in students' linguistic behavior online and in text messages. Fox 2007 studied gender differences in instant messaging and found that women sent messages that were more emotionally expressive than those sent by men. For more on sex differences in language, see Ullman 2008.

41 *girls about people and relationships:* For more on how women and men

use language differently and talk about different topics, see Tannen 1990. Women use words that reflect social concerns; men refer to more concrete objects and impersonal topics. Newman 2008 examined over 14,000 text files and found that women used more words related to psychological and social processes. Men referred more to object properties and impersonal topics. For more research on gender differences in language use, see Tannen 1997, Leaper 2007, and Ullman 2008.

42 *about objects and impersonal topics:* Pennebaker 2004.

42 *talk much about personal topics:* Newman 2008.

43 *brain at all before puberty:* Burnett 2009 found that, unlike basic emotions (such as disgust and fear), social emotions (such as guilt and embarrassment) require the representation of another's mental states; and during this transition, teens activated a different brain area for social versus basic emotions, while adults or children did not. See McClure 2004 for more on responses in the teen versus the adult brain.

43 *for social approval or disapproval:* Klucharev 2009 found that conflict with peer-group opinion triggered a brain response in the RCZ. They conclude that social-group norms evoke conformity via activity of the RCZ and ventral striatum. Jocham 2009 found that when an action leads to an unfavorable outcome, behavior needs to be adjusted and that in the human brain the RCZ is particularly responsive to performance errors and social disapproval. For more on the brain and social approval, see Tzur 2009, Yaniv 2009, and Behrns 2009.

43 *process of a massive recalibration:* For more on the brain effects of social exclusion and the distress of peer rejection during adolescence, see Masten 2009. For more on brain recalibration due to social emotions, such as embarrassment and guilt, see Burnett 2009.

43 *by their clans or tribes:* Freeman 2009a found culture shapes the brain's response.

43 *score and regained some respect:* Stanton 2009 found that high testosterone levels are associated with dominance behavior and pursuit of status in men, and that men's testosterone levels rise after winning dominance contests. This positive feedback to the brain primes future dominance behavior. For more on hormones and social status in males, see Sapolsky 1986 and 2005, Becker 2008b, Hermans 2006 and 2007 and 2008, Rubinow 2005, Van Honk 2005 and 2007, and Viau 2002.

43 *in front of his peers:* For more on the brain, social value, social learning, and self-confidence, see Behrens 2008 and Eme 2007.

43 *establish and maintain social hierarchy:* For more on the brain, social

hierarchy, dominance, and subordination, see Freeman 2009. For more on teen boys' behavior and hormones, see Olweus 1988 and Archer 2006.

44 *and not afraid to fight:* Sell 2009 found that males begin to show displays of hostile intent, such as angry facial expressions, in the teen years, and that males learn to quickly assess the strength and fighting spirit of other males just by looking at their faces.

45 *become hard to live with:* Olweus 1980 and 1988 found that teen boys have increased irritability.

45 *his team couldn't lose:* See Becker 2008a for more on the male brain and sex differences in excitement and dopamine systems. Salvador 1987, 2003, and 2005 found that during competition, a male's testosterone increases and, depending on the outcome and the importance of the event for the male, remains high for winners and drops for losers. Suay 1999 studied judo competitors. In those athletes larger increases in testosterone were highly correlated with looking angry while fighting, responding to a challenge, and being a violent competitor. For more on competition, brain, and testosterone, see Gatzke-Kopp 2009, Kahnt 2009, Sallet 2009, Kraemer 2004, and Berman 1993.

45 *losing, even in sports spectators:* Bernhardt 1998 found that even the vicarious experience of winning—e.g., being a fan whose team wins—leads to increased testosterone levels.

46 *in favor of their own:* Weisfeld 1999. See Levinson 1979 for more on psychological stages of development in adolescent males.

46 *seek autonomy from his parents:* Weisfeld 1999 and 2003. Fischer 2007 found that a high level of gender-role conflict in adult men was associated with parents' overprotection in teen years.

46 *they strike out on their own with bravado:* Spear 2004.

47 *new ideas in every generation:* Spear 2004 and Nelson 2005.

47 *willing to do risky things:* Nelson 2005. Steinberg 2007 found that adolescents and college-age individuals take more risks than children or adults do, which is reflected in statistics on automobile crashes, binge drinking, contraceptive use, and crime.

47 *consequences of unsafe, impulsive choices:* Steinberg 2007. Teicher 2000 found that the part of the brain that allows and encourages us to delay gratification and inhibit impulsive action—the PFC—won't be finished until later in the teen years and that it develops even later in boys' brains than in girls'.

47 *in a video driving game:* Steinberg 2004 found that the presence of peers more than doubled the number of risks teenagers took in a video driving game. Dahl 2008 says that sleep deprivation is rampant among adoles-

cents and that the consequences of insufficient sleep (sleepiness, lapses in attention, susceptibility to aggression, and synergy with alcohol) appear to contribute significantly to driving risks in teens.

47 *know what they're doing:* Eaton 2008 found that in the United States, 72 percent of all deaths among persons age 10–24 result from four causes: motor-vehicle crashes, other unintentional injuries, homicide, and suicide. The 2007 national Youth Risk Behavior Survey (YRBS) indicated that many high-school students engaged in behaviors that increased their likelihood of death from these four causes.

47 *biologically ready to handle independence:* Doremus-Fitzwater 2010 found that biological changes in the brain's motivational and reward-related regions increase teens' peer-directed social interactions, risk taking, novelty seeking, and drug and alcohol use relative to adults. For more on how teens' sleep deprivation influences risky peer interactions, see Dahl 2008.

48 *prefrontal cortex (PFC)—is like a brake:* Steinberg 2004 and 2007.

48 *boys until their early twenties:* Giedd 1996 and 2009.

48 *mother's body, but also by her smell:* Savic 2001 and Weisfeld 2003 found that the odor of genetically related family members is not romantically attractive. For more on MHC genes and odorous attraction, see Garver-Apgar 2006, Wedekind 1995, and Yamazaki 2007.

49 *to have fleeting sexual fantasies:* Campbell 2005 studied schoolboys age 12–18. The study found spontaneous nocturnal emission, secondary sexual characteristics, and salivary testosterone correlated with age at first sexual fantasies, noncoital sexual behavior, and coitus. They found first erection at an average age of 10.75 years, first sexual fantasy at 12.66 years, first spontaneous nocturnal emission at 13.02 years, and adult levels of blood testosterone at 17.2 years. Carlier 1985 found that boys' testicle size correlated best with first ejaculation.

49 *when boys begin frequent masturbation:* Korkmaz 2008 found that over 90 percent of 16-year-old boys masturbated and 98 percent reported that they liked it and thought it was natural. However, some felt guilt, fear of harming one's body, or shame. Some boys felt sexually inadequate when they compared themselves with their peers, feeling they were not as attractive to girls. For more on sexual behavior in teen boys, see Giles 2006, Auslander 2005, and Browning 2000.

49 *to three times a day:* Tanagho 2000.

49 *less than one time per day:* Korkmaz 2008. Gerressu 2008a found that 95 percent of men and 71 percent of women masturbated. And one consistency across all studies was the large gender difference in the prevalence and fre-

quency of masturbation—both being much greater in males. For more on sexual frequency in males and females, see Kontula 2002, Hyde 2005, Dekker 2002, Pinkerton 2002, Langstrom 2006, Giles 2006, and Laumann 1999b.

49 *at the first opportunity:* Tanagho 2000.

50 *finally came to "do it":* Adolescents are faced with many developmental tasks related to sexuality, such as forming romantic relationships and developing their sexual identities. Almost half will engage in vaginal sexual intercourse by the end of high school. For more on puberty and sexual development, see Eaton 2008 and Auslander 2005.

THREE: THE MATING BRAIN: LOVE AND LUST

51 *The Mating Brain: Love and Lust:* For a review on similarities and differences in the mate preferences and choices of women and men, see Geary 2004 and Young 2008.

52 *lit up like a slot machine:* For more on brain circuits for male courtship behaviors, see Pfaff 2002, Fernandez-Guasti 2000, Wu 2009, Maner 2007b, and Manoli 2006.

52 *of his ancient mating brain:* Voraceck 2006 found that the ultimate goal of female physical attractiveness is to elicit male sexual arousal. They found that males focused more on waist-to-hip ratio in women they saw moving and more on bust size in women who were stationary. In humans, the nucleus of the preoptic area of the hypothalamus is two to two and a half times larger in the male brain compared with the female. And Welling 2008 found that changes in testosterone levels contribute to the strength of men's attraction to femininity in women.

52 *in men across all cultures:* Singh 2002.

52 *mate-detection circuit was visual:* Tsujimura 2009 found that in the nonintercourse video clip, gaze time for the face and body of the actress was significantly longer among men than among women.

53 *mating brain read Nicole right:* Amador 2005 found that both sexes place high value on traits like dependable character, emotional stability/maturity, and pleasing disposition, as well as mutual attraction and love. Women in the study placed higher emphasis on ambitious/industrious character, similar educational background, and good financial prospects. Men cared more about the woman's fitness and good health, good cooking/housekeeping skills, and good looks.

53 *cute and looked harmless enough:* Maner 2008 and Shoup 2008. For more on mating judgments of men versus women, see Gangestad 1993 and 2000.

53 *cues as Ryan and Nicole:* Eibl-Eibesfeldt 1972.

53 *not out of his league:* Bateson 2005 and Alpern 2005 found that both men and women become less choosy over time, as the highest-ranking and most fit and attractive individuals pair off first. This means that for a male to be chosen by her as the best, all he has to do is outshine the other males he is being compared with on the specific characteristics she is focusing on.

54 *"I'm here to watch you":* O'Hair 1987 and Farrow 2003.

54 *wrong with meeting those expectations:* O'Hair 1987 and Haselton 2005.

55 *with the highest-pitched voices:* Apicella 2009. Sokhi 2005 found that female and male voices activate different regions of the male brain. Hughes 2008 and Pipitone 2008 found that a woman's voice attractiveness varies across the menstrual cycle.

55 *go weak in the knees:* And Roney 2008 found that women are more attracted to men's masculine characteristics, like jutting jaw and large muscles, during the ovulatory phase of the menstrual cycle.

55 *potentially a good genetic match:* Savic 2001b found that men are most attracted to the scent of women who are genetically different from them. And according to Lundstrom 2006, women who are on the "pill" or hormonal birth control do not make the same pheromones or have the ovulatory-phase rise in testosterone derivatives like androstenedione that stimulate the sweat glands to produce those feminine "come-hither" pheromones.

55 *give birth to sickly offspring:* Alvarez 2009.

55 *best to each other:* Wedekind 1995. Yamazaki 2007 found that dissimilar gene-type MHC emits special body odors that underlie mate choice and familial recognition, which helps make sure that inbreeding with parents and siblings does not occur. For more on brain responses to pheromones, see Hummer 2010, Mujica-Parodi 2009, and Prehn-Kristensen 2009.

55 *and not even known why:* Li 2007 says that people's odors can increase or decrease their likeability rating. Furthermore, according to Berglund 2006 and Sergeant 2007, gay males' and gay females' brains respond positively to same-sex pheromones. They dislike those odors of the opposite sex. For more on mating and pheromones, see Savic 2001a and 2009 and Zaviacic 2009.

55 *about hygiene; it's about genes:* Weisfeld 2003 and Olsson 2006. Havlicek 2009b found that olfactory and visual channels may work in a complementary way in mate attraction to achieve an optimal level of genetic variability.

56 *could talk to her about:* Keverne 2007 suggests that a male's ability to find a fertile mate requires some serious strategic maneuvering and that for

human males, these reproductive strategies are complex and embedded in the social structure and hierarchies of society. So success in human males usually depends more on intelligent behavior than on hormones or odors.

56 *tension between them was palpable:* Roney 2007 found that a man's testosterone goes up just from talking to a woman.

56 *secretly sent to their brains:* Gallup 2008 and Hughes 2007 found that kissing is a mate-assessment device. Wyart 2007 found that testosterone and its metabolites are found in male saliva, semen, and sweat—and they smell, and perhaps taste, delicious to a woman when she is ovulating. While females find the masculine odor attractive, heterosexual males dislike it. For more on pheromones and mating, see Bensafi 2003 and Walter 2008.

56 *center in a woman's brain:* Muir 2008 suggests that a man's excretions could be absorbed by a woman during kissing, touching, and skin-to-skin body contact and thereby affect her brain.

57 *Nicole was being so cautious:* Hill 2002 found women more cautious about moving toward sex too soon and men needing less of a sense of emotional investment in the relationship before having sex.

57 *up to three times longer:* Buss 1993.

57 *anything, least of all sex:* Roese 2006 found that women regret having sex early in the dating relationship more than men do.

57 *offspring he's likely to have:* Buss 1993 found that men's wanting sex with many women has likely been evolutionarily selected for in males.

57 *of education or financial independence:* Buss 2005 and Jensen-Campbell 1995.

57 *and was willing to invest:* Griskevicius 2007 found specific mating goals increased men's willingness to spend money on conspicuous luxuries for women. They say that romantic motives produce highly strategic and sex-specific self-presentations. And Klapwijk 2009 found that generosity serves the important purpose of communicating "trust."

58 *males who bring them meat:* Gomes 2009.

58 *men, sex often comes first:* For more on gender differences in love, commitment, and sex, see Roese 2006, Sprecher 2002, Keverne 2007, Loving 2009, McCall 2007, Geary 2000, and Buss 1993.

59 *side-blotched lizard* (Uta stansburiana)*:* Bleay 2007.

59 *"How can I pick a blue-throat?":* Humans have a mating system scientists refer to as mild polygyny—multiple partners combined with a variable

commitment to male parenting. Andrews 2008 found a sex difference in detecting infidelity—men are better at it. Atkins 2001a found that 20 percent to 25 percent of the married American population had had episodes of infidelity. Kontula 1994 found that in Finland, 52 percent of the men and 29 percent of the women reported episodes of infidelity in their lifetimes. They found that men reported being less emotionally involved than women with their infidelity partners, whereas the women seemed to connect both emotionally and sexually.

60 *aggressively reject all other females:* Gobrogge 2007 found that passionate mating changes the male brain biologically forever and that it leads pair-bonded male prairie voles to reject new fertile females. They found that it is an interaction between dopamine and vasopressin that results in pair bonding in the male brain's hypothalamus and nucleus accumbens, NAc. For more on hormones, genes, and pair bonding, see Winslow 1993, Carter 1998, Liu 2001 and 2003, Lim 2004c, and Young 2009a. (In the female brain, it is oxytocin and dopamine that interact to produce pair-bond formation.)

60 *preference for this one female:* For more on sex and partner preference in mammals, see Carter 1998 and Young 2008.

60 *bond with their sexual partners:* Liu 2001 found that when experimenters gave a chemical to block the vasopressin receptor, it blocked the intercourse-induced pair bonding in the males.

60 *in their brains couldn't merge:* For more on hormones in the brain, sex, and pair bonding, see Young 2008 and 2009b, Carter 1998, Becker 2008b, Wang Z. 2004, and Pfaff 2002.

60 *vole, he, too, became monogamous:* Lim 2004c experimentally induced pair-bond formation in the promiscuous vole by inserting the vasopressin gene from the monogamous prairie vole into the promiscuous vole.

60 *this vasopressin receptor gene too:* For more on the vasopressin receptor gene in humans, see Aragona 2009, Adkins-Regan 2009, and Walum 2008.

60 *to one woman for life:* Walum 2008 found an association between one of the human vasopressin receptor genes and traits reflecting pair-bonding behavior in men. They showed that the vasopressin genotype of men also affects marital quality as perceived by their wives.

61 *mating strategy for short-term partners:* Haselton 2005.

61 *to have sex with them:* Haselton 2005.

61 *and business and social connections:* Reviewed in Shackelford 2005d and Buss 2005b.

61 *and brain closer to Frank:* For more on female brain, oxytocin, and pair bonding, see Liu 2003.

62 *the more he squirmed:* Loving 2009 found an increased stress reaction in men when discussing commitment or marriage.

62 *electrical strain while they lied:* O'Hair 1987.

62 *couldn't get enough of her:* Gillath 2008a found that sex increases the desire to share personal information, fosters intimacy-related thoughts, and promotes a willingness to sacrifice for one's partner. Klusmann 2002 found that although sexual activity and sexual satisfaction decline in women and men as the duration of the partnership increases, sexual desire declines only in women, not in men. (And the desire for tenderness declines in men and rises in women.) They conclude that a stable pair bonding does not require high levels of sexual desire for women, after an initial phase of infatuation has passed. But for men the opposite is true. They found that male sexual desire should stay at a high level because it was selected for in evolutionary history as a precaution against the risk of sperm competition.

62 *necessary part of getting there:* For more on the male brain, pair-bond formation, and intercourse, see Liu 2001.

62 *getting a primitive biological craving:* For more on the specific brain areas where dopamine exerts its effects on pair bonding, pleasure, reward, and motivation, see Curtis 2006.

63 *neurotransmitter for motivation and reward:* Aragona 2009 found that dopamine transmission mediates the formation and maintenance of monogamous pair bonds. For more on motivation and reward in pair-bond formation, see Kruger 1998, Exton 2001a, and Young 2009.

63 *anticipation of pleasure and reward:* Knutson 2008 found that the nucleus accumbens (NAc) activation increases during anticipation of pleasure and deactivates during anticipation of loss in a relationship.

63 *mixed with estrogen and oxytocin:* Both males and females have oxytocin, vasopressin, testosterone, and estrogen, but the ratios are sex-specific and controlled by genes, proteins, and enzymes like aromatase. For more on the male brain, estrogen, and aromatase, see Wu 2009. For more on pair-bond formation in males and females, see Liu 2003, Bocklandt 2007, Becker 2008a, and Carter 2008.

63 *head over heels in love:* For more on the brain and intense romantic love, see Aron 2005 and Fisher 2005 and 2006.

63 *their bodies and brains became:* Gonzaga 2006.

63 *moments daydreaming about their lovers:* Fisher 2004.

63 *they focused only on Nicole:* Fisher 2006 found that when the in-love subjects looked at their beloveds, men also showed positive activity in a brain region associated with erection hardness. This means that the male love-response directly links romantic passion with a brain region associated with sexual arousal. Beauregard 2009 describes specific brain circuits for unconditional love.

64 *to hold on to her:* Buss 2002 says that the male must be fending off potential mate poachers and preventing his mate from defecting to hold on to the female. He found that mate-guarding adaptations evolved to avoid suffering negative reproductive costs, ranging from genetic cuckoldry to reputational damage to the permanent loss of a mate, and that male mate-guarding behaviors can range from vigilance to violence.

64 *happen several times a day:* For more on the male brain, lust, and visual sexual attraction circuits, see Fisher 2002, 2005, and 2006.

65 *imagined Frank hitting on her:* Rilling 2004 reviews sexual jealousy in males. Little 2007 and Burriss 2006 found that men sense a preference shift in their female partners toward more masculine men at ovulation. For more on male sexual coercion, see Starratt 2008 and 2007.

65 *a tactic called mate-poaching:* Schmitt 2004 found that the patterns for men in mate-poaching are similar across fifty-three nations. And they found that women poach too. Parker 2009 found that when a man was described as "unattached," 59 percent of the single women were interested in pursuing him, but when that same man was described as "being in a committed relationship," over 90 percent of the women expressed interest in the guy.

66 *intensify our feelings of love:* For more on rejection intensifying emotional commitment, see Baumeister 2001, Eisenberger 2004, Macdonald 2005, and Fisher 2002.

66 *and possessive mating instincts wild:* For more on mating instincts and hormones, see Carter 2007 and 2008, Becker 2009, and Pfaff 2002.

FOUR: THE BRAIN BELOW THE BELT

67 *average of one or two:* For more on men wanting a greater number of short-term sexual partners, see Schmitt 2001.

67 *interest in one-night stands:* For more on one-night stands, see Schmitt 2001, Laumann 1999b, and Mulhall 2008a. For more on men's satisfaction with their sex life, see Colson 2006, who found that almost 70 percent of men reported that they wished to change some things about their sexual life.

68 *men's testosterone levels to go up:* Van der Meij 2008.

68 *as sexually hot—or not:* Ortigue 2008 found that the male brain's decision about desirability of sexual stimuli occurs within the first 200 milliseconds after seeing a woman. This means it occurs before conscious processing.

69 *way their penis is shaped:* Sanchez 2007.

69 *happy with their partner's size:* Lever 2006 surveyed 52,031 men and women and found that many men wished they had a larger penis. And only 2 out of 1,000 men wished their penis were smaller. Dillon 2008 found that penile size is a considerable concern for many men from teens to old age. Wessells 1996 found that neither a man's age nor the size of his flaccid penis accurately predicted erectile length. But stretched penis length most closely correlated with erect penis length. For more on penis size, see Francken 2002.

69 *is their most important feature:* Francken 2002 found that a great many men believe that the size of the penis is directly proportional to its sexual power.

69 *larger than it needs to be:* Diamond 1997 notes that since the penis only has to be able to fit into a woman's vagina, men with penises that are too large may not be able to sire as many offspring, thus making larger penises undesirable.

69 *from 5.5 to 6.2 inches:* Wylie 2007 found that the average erect penis is 5.5 to 6.2 inches long and that an average-sized man is likely to be troubled by concerns that his penis is not large enough to satisfy his partner or himself and to be ashamed to have others view his penis, especially in the flaccid state.

69 *to their females, it's supersize:* Diamond 1997 describes that, compared with other mammals, the human penis is larger than necessary.

69 *conscious desire to have sex:* Janssen 2008 found in his survey that most men say that they can experience erections without feeling aroused or interested.

70 *to start an erection:* Tsujimura 2006. Holstege 2003 found that erection starts as a man imagines having sex with his partner or with other women in any of various positions and locations, both indoors and out. For more on erection, see Janssen 2008, Baskerville 2008, and Schober 2007.

70 *"order for him to function":* Beach 1967 found that none of this circuitry for sexual arousal or erection works in males who are deprived of testosterone. Steers 2000 found that it is testosterone, along with oxytocin and neurochemicals like dopamine, acetylcholine, and nitric oxide, that acts within the brain, spinal cord, and penis to produce an erection. Swann 2003 found that in the male brain, there is a sexually differentiated, testosterone-responsive network that relays signals to the muscle-control areas to produce copulation. For more on intercourse, see Redoute 2005.

70 *men to become fully erect:* Miyagawa 2007.

70 *the hormonal engines for erection:* Mouras 2008 found that while being shown sexual video clips, 8 out of 10 healthy men registered an erection, as demonstrated by a measuring cuff around the penis.

71 *hope of a sexual reward:* The nucleus accumbens (NAc) is a major center in the brain for anticipation of reward. For more on sex and reward, see Ponseti 2009 and Paredes 2009.

71 *"pay total attention to this now" circuits:* Lee 2006 and Moulier 2006.

71 *one smooth thrust, he was inside:* For more on vaginal penetration and condom use, see Crosby 2007.

72 *sexual tension, arousal, and pleasure:* Arnow 2002 and Holstege 2003.

72 *and fellatio twice as often as women:* Laumann 1999b.

72 *becomes less and less sensitive:* Payne 2007. Shafik 2007 found that stimulating the urethral opening keeps the nerves and muscles of the penis activated to maintain a throbbing, hard erection, thus enabling a forcefully ejected stream of semen, which has a better chance of impregnating the woman.

72 *from pain during sexual intercourse:* Payne 2007.

73 *neurochemical stars need to align:* Murstein 1998 found that men score higher than women in studies of sexual interest, frequency of sexual arousal, and sexual enjoyment.

73 *happens three minutes before entry:* Meston 2004. For more on women's orgasm, foreplay, and vaginal-penile intercourse, see Weiss 2009. For more on female sexual function and dysfunction, see Basson 2005.

73 *penis or clitoris to orgasm:* Georgiadis 2009. Muehlenhard 2009 found that both men and women pretend or fake orgasm—females 67 percent and men 28 percent of the time during penile-vaginal intercourse.

73 *periaqueductal gray (PAG)—activated intensely:* Parra-Gamez 2009 and Georgiadis 2009 found that the only prominent gender difference during orgasm was greater male activation of the PAG—the area for decreased pain and sexual moaning. Holstege 2003 found that brain scans taken of men while they are ejaculating show vivid activations in the ventral tegmental area (VTA), where dopamine is made.

74 *problems of his early twenties:* Revicki 2008 found that up to 75 percent of men ejaculate within ten minutes of penetration. Richters 2006 found that men had an orgasm in 95 percent of sexual encounters and women in 69 percent. Weiss 2009 found that women's likelihood of orgasm during intercourse

increases when penile-vaginal penetration lasts longer. For more on female sexual function, see Meston 2004, McKenna 2000, Mong 2003, and Basson 2005.

74 *or off by the brain:* Truitt 2002. Beaureguard 2009 found that the anterior cingulate cortex (ACC), the worrywart center, is alerted to the impending erection. This heads-up in the ACC lets it team up with other brain circuits, like the insula, or disgust center, to turn off the spinal erection generators when necessary.

74 *seven to thirteen minutes or more:* Waldinger 2005. Corty 2008 found that the normal, average length of intercourse is 3 to 13 minutes. Sex therapists recommend that men use Kegel squeezing exercises, masturbation, and mental distraction during intercourse, or condoms and penile-numbing gel if necessary, to treat rapid ejaculation. SSRI medication is also available to slow down ejaculation and help males last longer. However, SSRIs can prevent sexual arousal entirely.

74 *experienced it at least once:* Symonds 2007 and Revicki 2008. The diagnosis of PME, premature ejaculation, is made only when a lack of ejaculatory control interferes with sexual or emotional well-being in one or both partners. For more on the treatment of PME, see Sadeghi-Nejad 2008.

76 *achieving an erection:* Tanagho 2000 found that when the penis is massaged or when a sexual fantasy occurs in the brain, an erection is initiated by the PNS, the parasympathetic division of the autonomic nervous system. These PNS nerve branches cause the release of nitric oxide in the penis, dilating the arteries to fill it with blood and become hard. Viagra-like medicines act on the nitric oxide system to aid erections. Erection stops when the parasympathetic stimulation is discontinued and the SNS, the sympathetic division of the autonomic nervous system, starts constriction of the penile arteries, forcing blood out and making the penis get soft.

76 *get the erection he wanted:* Tanagho 2000. For more on erections, see Brody 2009 and Costa 2009.

77 *triggering the brain's sleep center:* Veenema 2008 found oxytocin is released in the male brain, during and after sex, for up to four hours, thus increasing sedation and relaxation and decreasing anxiety. For more on oxytocin and sex, see Waldherr 2007.

FIVE: THE DADDY BRAIN

79 *they're going to be fathers:* Buist 2003. Morse 2000 studied 327 healthy couples and found that some dads-to-be reported being distressed during the pregnancy about the relationship, performance failure at work, and/or sex.

80 *"the rest of its life":* Ahern 2009 and Meaney 2005. Boyce 2007 found that fathers who had insufficient information about pregnancy and childbirth were at risk of being distressed, suggesting that more attention needs to be paid to providing information to men about their partner's pregnancy, childbirth, and issues relating to caring for a newborn infant.

80 *down and prolactin goes up:* Gray 2006. Exton 2001b found that increased prolactin in males reduces sex drive. They suggest that prolactin goes up and testosterone goes down in dads-to-be in order to reduce sexual interest at a time when fertilization is not possible. For more on male hormones in pregnancy, see Delahunty 2007, Ma 2005, Burnham 2003, Wynne-Edwards 2000 and 2001, Carlson 2006, and Fleming 2002.

80 *from the mother-to-be's skin and sweat glands:* Vaglio 2009 found that during pregnancy, women develop a distinctive pattern of five volatile, odorous compounds in their sweat glands and skin that may act as pheromones.

80 *"sympathetic pregnancy":* Klein 1991 found that couvade is a common but poorly understood phenomenon whereby the expectant father experiences physical symptoms during his mate's pregnancy, like indigestion, increased or decreased appetite, weight gain, diarrhea or constipation, and even headache and toothache. Ziegler 2006 found that couvade occurs in other dad-to-be primates. In the study, expectant males showed significant increases in weight during the pregnancy, whereas control males did not. For more on couvade, see Conner 1990.

81 *growth of maternal brain circuits:* Larsen 2008 found that in female mice, contact with male pheromones caused new brain cells to grow in the female brain's frontal lobe in an area for maternal behavior. The data suggest that male pheromones stimulated a prolactin-mediated increase in brain cell growth in female mice, resulting in enhanced maternal behavior. For more on female brain and maternal circuits, see Becker 2008a.

81 *the three weeks before birth:* Storey 2000.

81 *to crying babies than non-dads:* Gray 2007.

81 *fathers give very little care:* Muller 2009. See Winking 2009 for fathering care in Bolivian men.

82 *calming them and promoting bonding:* For more on skin-to-skin contact between baby and parent, see Erlandsson 2007.

83 *fell in love with Blake:* For more on the brain and parental love, see Swain 2007, Feygin 2006, and Leckman 2004.

83 *each other's eyes and faces:* For more on the brain and parenting, see Leckman 2004.

83 *a seventh of a second:* Kringelbach 2008.

83 *before the dad feels compelled:* For more on parental responses to crying, see Bos 2010 and Fleming 2002.

84 *as he heard Blake crying:* Swain 2007 and 2008.

84 *day, for the first month:* Kozorovitskiy 2006. Also see Kinsley 2008 and Fleming 1999.

84 *hormones: prolactin, oxytocin, and vasopressin:* Kozorovitskiy 2006. Also see Berg 2001, Proverbio 2006, and Kuzawa 2009.

85 *child also activates the PFC:* See Kozorovitskiy 2006 for overview.

85 *studies by Dr. Ruth Feldman:* Feldman 2002 and 2007.

85 *difference between Mommy and Daddy:* Bretherton 2005. McElwain 2007 found that children (especially boys) benefit when parents differ in their reactions to their kids' emotions.

86 *mother's bond with her baby:* Matthiesen 2001 found that periods of increased massage-like hand movements or sucking of the mother's breast were followed by an increase in maternal oxytocin levels.

86 *differently to Mom and Dad:* Feldman 2003 and 2007.

86 *games Tim played with him:* Feldman 2007.

87 *playtime was much more spontaneous:* Cannon 2008.

87 *were in the driver's seat:* Schoppe-Sullivan 2008 found that moms are ultimately the gatekeepers for the dads' access to kids. They also found that fathers who are in more harmonious marriages are more affectionate toward their infants. For more on marriage and parenting, see Fagan 2009.

87 *could look to for help:* Silk 2009 found that the help females receive from their own mothers and adult daughters—and other female kin—has a significant influence on children's survival and well-being. For more on parenting, female kin, and child survival, see Kendler 2005, Taylor 2000, Hill 2003, Hawkes 2004, Sear 2008, and Gurven 2009.

88 *it comes to staying together:* Pasley 2002 found that dads who perceived their wives as evaluating them positively as fathers were likely to have more commitment to the marital relationship. For more on marital relationship and parenting, see Roopnarine 2005.

88 *improves their ability to learn:* Feldman 2007.

88 *unpredictable and thus more stimulating:* For more on fathers stimulat-

ing their children, see Pecheux 1994, O'Neill 2001, Fernald 1989, Grossman 2002, and Pancsofar 2008.

88 *dads were more quirky and fun:* O'Neill 2001 and Pecheux 1994.

89 *the time they reached adolescence:* Grossmann 2002 found that in their sixteen-year longitudinal study, fathers' sensitive and challenging play was a key variable in a child's success. For more on children's success and fathering, see Sarkadi 2008.

89 *recognize mental tricks and deceits:* Bretherton 2005.

90 *more direct orders than mothers do:* Abkarian 2003.

90 *need the way Mom does:* Fernald 1989.

90 *especially sons, toe the line:* Sarkadi 2008.

91 *the hormones testosterone and vasopressin:* Wang Z. 1993 found that having no testosterone at all, due to castration, reduces paternal behaviors. Brain connections in castrated male mammals for paternal behavior became reduced due to fewer vasopressin cells in the brain.

91 *be better, more protective dads:* Frazier 2006. For more on brain motivation and paternal care, see Devries 2009 and Becker 2009.

91 *dads who were not disciplinarians:* Sarkadi 2008.

92 *with men later in life:* Wiszewska 2007.

92 *always more negotiating and compromising:* Tannen 1995. Leaper 1998 found that in studies of two parents with their children, there was greater talkativeness of mothers to their children than of fathers.

93 *doing something to help them:* Bretherton 2005.

94 *their sons are very young:* Leckman 2004 and Feldman 2002.

94 *high parental care in childhood:* Pruessner 2004.

94 *bonding parent to child:* Feldman 2002.

SIX: MANHOOD: THE EMOTIONAL LIVES OF MEN

96 *parts of our basic biology:* Kozorovitskiy 2005.

97 *temporal-parietal junction system, or TPJ:* Shamay-Tsoory 2009 found two systems for empathy: one for cognitive empathy, the other for emotional empathy. Schulte-Ruther 2008 found gender differences in brain networks for empathy. The TPJ is a hub where many circuits for attributing mental states to others converge and diverge, for example, the superior temporal sulcus, medial prefrontal cortex, and others. For more on hubs in the brain,

see Thioux 2008 and Immordino-Yang 2009. The MNS, the mirror neuron system, which is spread out through many brain regions in humans, helps us understand how others feel, how they act, and what they will do. Structures within the human mirror neuron system are found to be involved in sharing gestures and facial expressions. Zaki 2009 found that using these two sets of brain regions helps us accurately track the attributions we make about another's internal emotional state. Yuan 2009 found that men make less accurate judgments than women when subtle or moderately negative emotions are expressed, but are similar in accuracy when highly negative emotions are expressed.

97 *use the other system more:* Schulte-Ruther 2008 found increased neural activity in the TPJ, temporal-parietal junction, in males; and found that females showed increased activation of the MNS, specifically the inferior frontal mirror neurons. Thus, females recruited areas containing mirror neurons to a higher degree than males during emotional processing in empathetic face-to-face interactions. Witelson 1991a found that the temporal-parietal region, TPJ, of the brain is larger in males. Cheng 2009 found that females perform better in empathy, interpersonal sensitivity, and emotional recognition than do males, perhaps since the mirror neuron system, MNS, plays an important role in these processes. The researchers found that young adult females had significantly larger gray matter volume in the mirror neuron system than did males. Yuan 2009 hypothesizes that the larger female MNS may result in more emotional contagion—or infectious feelings—and empathy in the average female than in the average male.

97 *This is called emotional empathy:* Bastiaansen 2009. The capacity that humans have to intuitively grasp the mental states of other individuals is important for social functioning. Even when people's more subtle emotions remain puzzling, we can have gut feelings of what is going on in them. The MNS plays a major role in this skill.

97 *This is called cognitive empathy:* Cognitive empathy means intellectually understanding what is upsetting to the other person in front of us—but *not* feeling that same feeling in our gut. This mental separation of one's own perspective from that of another person helps us to disentangle our own feelings from those observed in other people and to figure out the solution to an emotional problem without becoming "infected" with their emotions. The TPJ hub is involved with doing this. For more on keeping one's own emotions separate and the TPJ in males, see Schulte-Ruther 2008.

97 *may cement a preference for it:* Christakou 2009. For more on gender differences in empathy, see Schulte-Ruther 2008, Becker 2008b, and Eme 2007.

97 *boundary between emotions of the "self":* Schulte-Ruther 2008 found that men's brains also have increased activation in the TPJ during the at-

tribution of emotion to themselves, thus keeping a boundary between self and other.

98 *off by a blank face:* Schulte-Ruther 2008 found that there are gender differences in facial mimicry. The study shows more involvement of the MNS in females than in males during empathy-related face-to-face interactions. For more on gender differences in facial mimicry, see Dimberg 1990.

98 *switches over to the TPJ:* Schulte-Ruther 2008.

98 *he did* feel *her distress:* Schulte-Ruther 2008.

98 *system and share her emotions:* Wild 2001. For more on gender differences in facial imitation and emotional contagion, see Sonnby-Borgström 2008. For a review on emotional contagion and emotional and cognitive empathy, see Nummenmaa 2008.

99 *the unwritten laws of masculinity:* Brod 1987.

99 *muscles to mask his fear:* The facial muscles can reflect what is going on inside our brains, so learning to hide fear must be practiced. Males train their faces in gender-specific ways, as do females. But the unconscious moment of feeling or recognizing fear or contempt cannot be completely covered up, especially in a brain scanner—the ultimate lie detector. For example, Aleman 2008 says that the male brain reacts more strongly than the female to signals of status or hierarchy, especially to a look of contempt, for that is the universal facial expression of superiority.

99 *emotionally reactive than the women:* The unconscious mind subliminally triggers the facial muscles during an emotion, if only for a few milliseconds. These facial expressions are called *microexpressions* and can be measured by hooking detectors up to the facial muscles. For more on facial muscles and expressions, see Ekman 1978. Sonnby-Borgström 2008 found gender differences in facial-muscle responses representing information-processing levels from subliminal (spontaneous/unconsious) to supraliminal (conscious/emotionally regulated). The researchers also found that men consciously (supraliminally) suppressed emotions, but unconsciously, at first, they reacted more to their emotions—as evidenced only by the microexpressions in their frowning or smiling muscles.

100 *subtle frown to a pout:* Sonnby-Borgström 2008 found that women consciously (supraliminally) exaggerated their emotions, but unconsciously, at first, they reacted less to their emotions. For more on sex differences in smiling, see Hecht 1998 and Weyers 2009.

100 *". . . with logic instead of feelings?":* For more on sex-related differences in brain activity during emotion regulation, see Mak 2009.

101 *to run on different hormones:* Holden 2004 and Eme 2007.

101 *for our different emotional styles:* For more on gender differences in emotional styles, see Eme 2007, Baron-Cohen 2004c, and Hines 2004.

101 *differently in men and women:* Baron-Cohen 2004c and Eme 2007.

101 *men became temporarily more empathetic:* Domes 2007 found that men's ability to infer the emotional and mental state of others improved after intranasal administration of oxytocin. For more on oxytocin, testosterone, and generosity, see Zak 2009. Barraza 2009 showed that oxytocin given to men increased empathy and generosity.

101 *made them more mentally focused:* Hermans 2008.

102 *than it is for women:* For more on male brain circuits for anger, aggression, and physical fighting, see Lindenfors 2007, Eme 2007, Dunbar 2007a, and Williams 2006.

102 *hormonally reinforced during the teen years:* Eme 2007. For more on hormones and angry expressions, see Wirth 2007.

102 *familiar part of his life:* Eme 2007. For more on social and physical risk-taking in males, see Xue 2009, Fuxjager 2009, Wirth 2007, Carre 2008, and Hand 2009.

102 *twenty times more often than women do:* Campbell 2006.

103 *firing up his fighting circuits:* Wirth 2007.

103 *and body for a fight:* Becker 2009. For more on testosterone changes and power motivation in victory versus defeat, see Schultheiss 2005. For more on anger and driving, see Leal 2008.

104 *changes prompting his aggressive behavior:* Stanton 2007 and 2009b found that higher testosterone makes the male amygdala less responsive to fear and more ready to fight in the face of a dominance challenge. For more on dominance, testosterone, and physical aggression, see Mazur 1998, Archer 2006, Eme 2007, and Carre 2008.

105 *emotional events better and longer:* Cahill 2004. Canli 2002 found that women had more brain regions where emotions enhanced memory more powerfully, causing women to remember emotional events better than men.

105 *for memory enhanced by emotion:* Phelps 2004 found that the amygdala and hippocampal complex are linked to two independent memory systems and that in emotional situations, these two systems interact in subtle but important ways.

105 *the emotion that they're feeling:* Canli 2002 and Cahill 2004.

105 *and activate his territorial fight reaction:* Stanton 2009b. Van Honk 2007 found that in humans a surge of testosterone reduces fear responses and stress-axis reactivity in the brain, altering the natural avoidance of threats by decreasing fear.

105 *their anger harder to control:* Wirth 2007 found that anger in faces was pleasurable and reinforcing to those with higher testosterone. They suggest that testosterone specifically encourages approaching and engaging with angry faces since it is a signal of dominance challenge. Carre 2009 found that changes in testosterone can ignite future aggressive behavior in men.

106 *angry got him fired up:* For more on testosterone fueling aggression, see Stanton 2009b, Wirth 2007, and Archer 2006.

106 *if those feelings are unpleasant:* Tamir 2008 found that individuals may choose to experience emotions that are instrumental (useful), despite short-term hedonic (unpleasant) costs.

106 *us think more clearly, too:* Tamir 2008.

106 *people* more *rational, not less:* Tamir 2008. Anger has been claimed to trigger superficial, nonanalytic information processing, but Moons 2007 found that induced anger promoted analytic processing. The study showed that angry people can have both the capacity and motivation to process and think more clearly.

106 *acceptable to express their anger:* For more on males' anger expression, see Dabbs 1996, Mazur 1998, and Archer 2006. Archer 2006 reports that the fear of males in females begins in the first or second grade. Archer 2009 found that sex differences in physical aggression increase with the degree of risk, occur early in life, peak in young adulthood, and are likely to be mediated by greater male impulsiveness and greater female fear of physical danger.

106 *more dramatically to being challenged:* Schultheiss 2003. For more on high-testosterone men, see Archer 2006.

107 *more aggressive than subordinate males:* For more on dominance and subordination in primates, see Wrangham 2004, Sapolsky 1986 and 2005, and Archer 2006.

107 *dominance, thus increasing his testosterone:* For more on dominance challenges, fear, and testosterone, see Van Honk 2001, Hermans 2006, and Josephs 2006. Mehta 2009 found that high-testosterone winners chose to repeat the competitive task, whereas high-testosterone losers chose to avoid it.

107 *by other men but also by women:* Williams 2006 found that there is a perceptual system in both males and females that has evolved to rapidly

detect aggression in males and that angry men get noticed more by both sexes. Cox 1999 found that angry men are also judged to be more competent; angry women are judged to be less competent. Roney 2006 found that women have a preference for male faces that indicate higher testosterone.

107 *"you suppressed your anger altogether":* Harburg 2008 found that a good fight with your spouse may keep you and your marriage healthier.

109 *tendency toward anger and aggression:* Maner 2007 found that men behave differently when the dominance hierarchy is unstable and there is potential for losing their power.

109 *dialed down by social conditions:* Reber 2008 found that subordinate/dominant behavior in an unstable hierarchy influences the hypothalamic production of vasopressin and that subordinate males' exposure to dominants results in diminished body weight and increases anxiety-related behavior. They found that the hormones of aggression in the brains of subordinates decrease after only twenty days.

109 *factors that dial it down:* Burnham 2003 found that testosterone is lower in men who are in stable, committed relationships.

109 *preparing them for turf wars:* Ferris 2008a found that the neural circuitry of aggression and territorial defense can be dialed up with vasopressin.

109 *in his hypothalamus and amygdala:* Ferris 2008a found that the hypothalamus and amygdala help bridge the emotional, motor, and cognitive components of the brain's aggressive responses. The study found that drugs that block vasopressin neurotransmission suppress activity in circuits for aggression and motivation.

109 *is wired into the male brain:* Kozorovitskiy 2004 found that differences in social status correspond to structural differences in the male brain. They found that higher social status accounts for the effect that dominance has on growing new brain cells in the adult male brain.

110 *to gain or maintain rank:* Wrangham 2004.

110 *as they anticipate a confrontation:* For more on rank, aggression, and confrontation in humans and primates, see Mazur 1998, Archer 2006, and Stanton 2009.

110 *opponents within their own species:* For more on male-male fighting and competition, see Motta 2009, Wrangham 2004, and Archer 2006.

110 *circuitry for this instinctive one-upmanship:* Motta 2009 found an area in the male brain's hypothalamus, called the DPN, is activated for instinctive one-upmanship in male rats for territory protection against higher-ranking males.

SEVEN: THE MATURE MALE BRAIN

112 *he had in his thirties:* McCrae 1996 found that personality is stable throughout life.

112 *suited to a luxury sedan:* Mehta 2009 found that men with lower testosterone are more motivated to cooperate with others. Mykletun 2006 found that men in their fifties are more satisfied with their sex lives than men in their thirties and forties. And that men in their fifties recorded similar levels of satisfaction as twenty- to twenty-nine-year-old men.

113 *make less testosterone and vasopressin:* Rosario 2004 and 2009. Geenen 1988 found that young men make more vasopressin than oxytocin due to their high testosterone levels. Debiec 2005 found that oxytocin increases feelings of attachment and love, whereas vasopressin increases feelings of territoriality and defense. Huber 2005 found that vasopressin and oxytocin modulate the brain's circuits for love and fear via their different effects on the amygdala.

113 *increases as men get older:* Rosario 2009 found that in male brains, during normal aging, there was no decrease of estrogen but there was a decrease of testosterone. The ratio of estrogen to testosterone in men thus increased with age. Berchtold 2008 found that clear gender differences in brain aging are evident, indicating that the brain undergoes sex-specific aging changes in gene expression, not only in the developmental period but also in later life.

113 *the cuddling and bonding hormone:* Domes 2007b. Ditzen 2009 found that the effects of increased oxytocin on a couple's relationship could be measured during conflict resolution. When researchers gave more oxytocin, it significantly increased positive communication behavior in relation to negative behavior during the couple's conflict discussion. Heinrichs 2008 says that vasopressin has been implicated as a primary factor in male-typical aggressive social behaviors, whereas oxytocin reduces conflict, anxiety, stress, and aggression. Thus relatively more oxytocin and less vasopressin may improve conflict resolution in close relationships.

113 *to read subtle facial expressions:* Domes 2007b found that oxytocin improved the men's performance on a test of the ability to "read the mind" of others from just looking at subtle facial expressions of the eyes and face. Unkelbach 2008 found that oxytocin increases the use of relationship words and language.

113 *more dramatic effect on them:* Juntti 2008 and Jordan 2008. Kosfeld 2005 found that giving young men oxytocin increased their ability to trust others. Fliers 1985 found that with aging, the male brain changes in areas producing vasopressin, especially in brain areas where vasopressin innervation is dependent on higher testosterone levels. Thus as testosterone goes

down and estrogen stays the same, the male brain may make not only less vasopressin but more oxytocin, becoming more like the female's.

114 *she looked up to him:* For more on the brain circuits involved in admiration, see Immordino-Yang 2009.

115 *listener, and be more affectionate:* Burri 2008 found that when oxytocin was increased in men, the men's sexual arousal increased. And when the men in the study were asked about their subjective perception of whether oxytocin or placebo had been applied, eight out of ten subjects in the oxytocin group answered correctly, thus pointing to an altered perception of sexual arousal by oxytocin.

116 *as his testosterone production declined:* Johnson 2006 found that higher testosterone increases the drive to outdo other men.

116 *no longer worth it to him:* Winning and losing may not matter so much, and cooperation may matter more to men as their testosterone levels mellow out. Mehta 2009 found that high-testosterone men are motivated to gain status (good performance in individual competition), whereas low-testosterone men are motivated to cooperate with others (good performance in intergroup competition). So with age, a man's need for dominance may have biologically subsided.

116 *group stability, and mediate conflict:* Yamagiwa 2001.

116 *up in trees for protection:* Yamagiwa 2001.

117 *meant more than being lovers:* For more on the hormones of pair bonding, see Young 2009 and Carter 1998. Kendrick 2000 found that after sexual intercourse, in the female the stimulation of the cervix and vagina, along with orgasm, causes a coordinated release of oxytocin in the brain that acts to increase maternal and pair-bonding behaviors lasting for up to one hour.

117 *she is no longer fertile:* Tuljapurkar 2007 found that female menopause, around age fifty, should theoretically be followed by a sharp increase in mortality for humans, a "wall of death." Their analysis showed that older men reproducing with younger women throughout evolution is what forms the basis for increasing human longevity in both sexes. For more on the genes of longevity, see Emery 2007.

118 *in men who are social:* Cacioppo 2009c.

119 *the long run, as smoking:* Cacioppo 2009b.

119 *engraved into their brain circuits:* It turns out that for men, health and marriage are more connected than for women because men get most of their social interaction after retirement via their wives' social connections and

social planning skills. For more on the brain and aging, see Decety 2009, Dedovic 2009, and Cacioppo 2009a.

119 *circuits are weakened from disuse:* Decety 2009 found that men need to use their social and emotional cognition and problem-solving circuits, or else their temporal-parietal junction (TPJ) will atrophy. The TPJ is associated with cognitively taking the perspective of another person, which happens only if other people are around to activate that part of the brain. Also see Cacioppo 2009b.

119 *retained their reproductive abilities longer:* Schmidt 2009.

120 *people who are socially isolated:* Cacioppo 2009b found that the brain's TPJ is much less activated among the lonely than in the nonlonely. And the lonely also activate their brains' reward circuits less. So they begin to get less pleasure out of interacting with others, and their brains' social circuits may become less responsive if they stay lonely too long.

120 *get married and stay married:* Willcox 2006 found that before age eighty-five, the lack of a marital partner is associated with increased mortality. Schmitt 2007 found that marital satisfaction is important for health and well-being and that a high quality of interpersonal interaction is particularly important in predicting marital satisfaction and long marriages.

120 *men live 1.7 years longer:* Smith 2009 found that marriage factors and health differ for men and women throughout the life span because being married shortens a woman's life by 1.4 years but lengthens a man's by 1.7 years. The study found that bad marriages, with lots of arguing and negative feelings, make both sexes stressed but cause physical illness only in wives. The men in bad marriages reported stress and reported being depressed, but their physical health didn't seem to be affected. For more on marriage and health, see Kiecolt-Glaser 1998, 2001, and 2005, Gabory 2009, Behar 2008, Willcox 2006, Felder 2006, and Levenson 1993.

121 *make a man a man:* Beach 1967 found that the male brain's sexual-pursuit and arousal circuits must be primed for action by testosterone in order for the man to function. For more on testosterone and sexual function, see Moffat 2007.

122 *so-called andropause, or male menopause:* Sharma 2009 found that the benefits of treating severely low testosterone are well established. Milder forms of low testosterone in the aging male, known as andropause, are common starting in the fifties and sixties. Researchers say that testosterone replacement therapy may produce a wide range of benefits for men, including improvement in libido, bone density, muscle mass, body composition, mood, and cognition. But they say there is no definitive verdict on the balance be-

tween risks and benefits of testosterone replacement therapy in aging males. For more on testosterone replacement in the aging male, see Lapauw 2008, Anawalt 2001, Qian 2000, Heaton 2001, Lincoln 2001, Hogervorst 2009, and Van Strien 2009.

122 *of life also includes sex:* Smith 2007 found that desire for sexual activity remained high among men seventy and older. Colson 2006 found that almost 70 percent of men reported they would like to change some things about their sex life. Kontula 2002 found that, especially with the increased duration of the relationship, the frequency of masturbation actually increased in men who were in long-term partnerships, even if penile-vaginal sex was occurring regularly. Beaulieu-Prevost 2007 found that men after age forty continue to have reflexive erections while they're sleeping. Eight percent of older men's dreams were about intercourse and resulted in nocturnal erection. Laumann 1999 found that masturbation increased in older men because their current partner did not want sex; this was three times more often in men than in women.

122 *at this stage in life:* Siegel 2007 found that sexual changes that occur in men as they get older affect their self-perception and sexual identity. Janssen 2008 found that changes in the quality of older men's erections had a direct effect on their sexual encounters, including, for some, a shifting of focus to the partner and her sexual enjoyment. In this study, older men said that as they aged, they became more careful and particular in choosing sexual partners.

123 *they produced in their twenties:* Tanagho 2000 found that a man's testosterone peaks at about age seventeen; then it plateaus at a high level for a while and may slowly start to slide by the thirties and forties; then, by eighty, his testosterone may be less than half of what it was when he was young. Vermeulen 1999 says that as age increases, a man's body fat increases and his testosterone goes down. For more on testosterone and aging, see Qian 2000, Araujo 2007, and Laughlin 2008. For approaches to evaluating and treating men with low testosterone, see Snyder 2008.

123 *quickly becoming a marital problem:* For comprehensive research data on age, couples, and sex, see Laumann 1999. Erectile dysfunction (ED) is a common condition estimated to affect more than 150 million men worldwide. Doctors now believe ED should be regarded as a shared sexual problem because it has significant detrimental effects both on the men who experience it and on their partners. So the wives or partners should be included in treatment decisions for Viagra-like drugs or testosterone. Isidori 2005a found that in older men with sexual complaints, testosterone treatment moderately improved the number of nocturnal erections, sexual thoughts and motivation, number of successful intercourses, scores of erectile function, and over-

all sexual satisfaction, but that in men with normal testosterone levels at the beginning, the testosterone had no effect on erectile function compared to placebo. Mulhall 2008b found that 74 percent of men were willing to take erection-enhancing drugs if necessary. For more on treatment, see Wang P. 2009 and Sharma 2009.

123 *organs working at full capacity:* Redoute 2005 found that the brains of men with very low testosterone did not light up in the areas necessary for sexual arousal as they watched sex movie clips in the brain scanner. But after the males were injected with testosterone, these brain areas flashed back on again.

124 *testosterone can kick-start abdominal weight loss:* Gooren 2009. Srinivas-Shankar 2009b found that testosterone treatment in older men with low levels of testosterone might have beneficial effects on body composition (losing fat), muscle strength, sexual function, and cognition.

124 *men, that's still not enough:* For more on DHEA, androgens, and aging, see Rainey 2008, Baker 2006, Dharia 2004, Anawalt 2001, Parker 1999, and Sapolsky 1993.

124 *results from androgen-replacement therapy:* Cherrier 2007 studied the cognitive responses of older men in whom moderate or large increases in serum testosterone levels were induced by testosterone supplementation, and found that those with moderate testosterone replacement did better on cognitive tests.

124 *and it's not for everyone:* Wang P. 2009 says men must be monitored closely for prostate function and for prostate and breast cancer when considering or being prescribed testosterone replacement.

124 *there may be other answers:* Exercise, lowered stress, loving relationships, and healthy diet are well-proven aids to healthful aging in both men and women. Also see Roberts 2008.

125 *men as the hormone DHEA:* Unpublished data. Soma 2008 found that the adrenal androgen precursor dehydroepiandrosterone (DHEA) may be important for the expression of sexuality and aggression when gonadal testosterone synthesis is low. DHEA is metabolized into active sex steroids, both in the periphery and within the brain itself. For more on DHEA in development and aging, see Parker 1999.

125 *marriage, even into old age:* Laumann 1999a.

125 *"asked my doctor for Viagra":* Rosen 2006 found that for mature men, Viagra started a revolution, and this was true even for older men who did not have ED (erectile dysfunction). This is because older men now get the message: "Other guys are having a lot of sex as they get older. Why not me?"

The researchers also found that the outcome in men with erectile dysfunction could be helped or hindered by their partners or wives.

126 *as much as threefold:* Zhang 2007 found a surprising result. These so-called impotence drugs not only block the PDE-5 enzyme to aid erection, but can also boost levels of oxytocin.

126 *beneficial reduction in blood pressure:* Holt-Lunstad 2008 found that the influence of a warm touch enhancement intervention among married couples improves blood pressure, oxytocin, alpha amylase, and cortisol. But only husbands in the intervention group had significantly lower post-treatment twenty-four-hour systolic blood pressure than the control group. McGlone 2007 found that soft touches also activate a class of slow unmylenated nerves that send messages back to the brain's insular cortex, limbic system, and orbitofrontal cortex. These nerves give a pleasant sensation from light touch, and when the skin is lightly stroked, the brain interprets it as emotional touch.

126 *for every one critical remark:* Gottman 2006 found that the marriages of couples having a videotaped conflict were more likely to survive if they made roughly five positive, complimentary comments for every one negative, critical comment. Carrere 1999 found that it was possible to predict a good or bad marital outcome over a six-year period using just the first three minutes of data. They found that couples who used a ratio of five positive comments to one negative (the Gottman technique) had marriages that survived and even improved.

127 *letting go of them:* St. Jacques 2009.

127 *"wisdom* does *come with age":* Mather 2005 found that as people get older, they experience fewer negative emotions. They say that this "positivity effect" in older adults' past memories, compared with younger people, seems to be implemented by cognitive control mechanisms that enhance positive and diminish negative thinking. For more on the positivity effect in older brains, see Ashley 2009, Charles 2008, Nielsen 2008, Dreher 2008, and Samanez-Larkin 2007.

128 *The Grandfather Brain:* Szinovacz 1998b found that grandparenthood is initially a dual process signifying both the transition to parenthood for the adult child and the transition to being a grandparent for his/her parents. For more on similarities in the responses of men and women to grandparenthood, see King 1998b.

128 *been when Ali was born:* Beauregard 2009 found that the neural basis of unconditional love shows that romantic love and parental love are mediated by regions specific to each, as well as overlapping regions in the brain's reward system.

129 *or the stage of generativity:* Vaillant 2002. Vaillant wanted to explore how men adapt to circumstances over a lifetime. The average age of the Harvard men is now eighty-seven. He has concluded that in these men, sustaining warm, close relationships turned out to be a powerful predictor of successful aging.

129 *food from old to young:* Kaplan 1997.

129 *love for their adult children:* Roberto 2001.

129 *their adult children, and grandchildren:* Kivett 1998.

129 *relationship between grandparents and grandchildren:* Silverstein 2001 and Szinovacz 1998b. Jiang 2007 found that grandparents were dominant in shaping children's eating behavior in three-generation families.

APPENDIX: THE MALE BRAIN AND SEXUAL ORIENTATION

133 *males as in straight males:* Swaab 1990.

133 *reacts with the developing brain:* Swaab 1995.

133 *males than in straight males:* Allen 1992.

134 *of the straight male brain:* Savic 2008.

134 *in gay and straight men:* LeVay 1991 and Kinnunen 2004.

134 *straight women on such tasks:* Rahman 2005 and 2008.

135 *to the face of a man:* Swaab 2009.

REFERENCES

Aarts, H., and J. van Honk (2009). "Testosterone and unconscious positive priming increase human motivation separately." *Neuroreport* 20(14): 1300–1303.

Abkarian, G. G., J. P. Dworkin, et al. (2003). "Fathers' Speech to Their Children: Perfect Pitch or Tin Ear?" *Fathering* 1(1): 27–50.

Abrahamson, D. (2004). "Embodied spatial articulation: A gesture perspective on student negotiation between kinesthetic schemas and epistemic forms in learning mathematics." Paper presented at the annual meeting of the North American Chapter of the International Group for the Psychology of Mathematics Education, Delta Chelsea Hotel, Toronto, Ontario, Canada.

Abrams, D., A. Rutland, et al. (2009). "Children's group nous: Understanding and applying peer exclusion within and between groups." *Child Dev* 80(1): 224–43.

Abrams, D., A. Rutland, et al. (2008). "Children's judgments of disloyal and immoral peer behavior: Subjective group dynamics in minimal intergroup contexts." *Child Dev* 79(2): 444–61.

Abrams, D., A. Rutland, et al. (2003). "The development of subjective group dynamics: Children's judgments of normative and deviant in-group and out-group individuals." *Child Dev* 74(6): 1840–56.

Achenbach, G. G., and C. T. Snowdon (2002). "Costs of caregiving: Weight loss in captive adult male cotton-top tamarins (*Saguinus oedipus*) following the birth of infants." *Int J Primatol* 23(1): 179–89.

Adkins-Regan, E. (2009). "Neuroendocrinology of social behavior." *ILAR J* 50(1): 5–14.

Aguiniga, D. M., C. Streeter, et al. (2007). "The XY-zone male involvement project: Guiding male teenagers as they journey into manhood." *Children & Schools* 29(2): 119–22.

Ahern, T. H., and L. J. Young (2009). "The impact of early life family structure on adult social attachment, alloparental behavior, and the neuropeptide systems regulating affiliative behaviors in the monogamous prairie vole (*Microtus ochrogaster*)." *Front Behav Neurosci* 3: 17.

Ahmed, E. I., J. L. Zehr, et al. (2008). "Pubertal hormones modulate the addition of new cells to sexually dimorphic brain regions." *Nat Neurosci* 11: 995–97.

Albrecht, L., and D. Styne (2007). "Laboratory testing of gonadal steroids in children." *Pediatr Endocrinol Rev* 5 Suppl 1: 599–607.

Aleman, A., and M. Swart (2008). "Sex differences in neural activation to facial expressions denoting contempt and disgust." *Public Library of Science One* 3(11): e3622.

Allen, L. S., and R. A. Gorski (1992). "Sexual orientation and the size of the anterior commissure in the human brain." *Proc Natl Acad Sci U S A* 89(15): 7199–7202.

Almond, R. E., T. E. Ziegler, et al. (2008). "Changes in prolactin and glucocorticoid levels in cotton-top tamarin fathers during their mate's pregnancy: The effect of infants and paternal experience." *Am J Primatol* 70(6): 560–65.

Alpern, S., and D. Reyniers (2005). "Strategic mating with common preferences." *J Theor Biol* 237(4): 337–54.

Alvarez, G., F. C. Ceballos, et al. (2009). "The role of inbreeding in the extinction of a European royal dynasty." *PLoS One* 4(4): e5174.

Amador, J., T. Charles, et al. (2005). "Sex and generational differences in desired characteristics in mate selection." *Psychological Reports* 96(1): 19–25.

Anawalt, B. D., and G. R. Merriam (2001). "Neuroendocrine aging in men: Andropause and somatopause." *Endocrinol Metab Clin North Am* 30(3): 647–69.

Andreano, J. M., and L. Cahill (2009). "Sex influences on the neurobiology of learning and memory." *Learn Mem* 16(4): 248–66.

Andreano, J. M., and L. Cahill (2006). "Glucocorticoid release and memory consolidation in men and women." *Psychol Sci* 17(6): 466–70.

Andrews, P. W., S. W. Gangestad, et al. (2008). "Sex differences in detecting sexual infidelity: Results of a maximum likelihood method for analyzing the sensitivity of sex differences to underreporting." *Human Nature* 19(4): 347–73.

Angelopoulou, R., G. Lavranos, et al. (2006). "Establishing sexual dimorphism in humans." *Coll Anthropol* 30(3): 653–58.

Apicella, C. L., and D. R. Feinberg (2009). "Voice pitch alters mate-choice-relevant perception in hunter-gatherers." *Proc Biol Sci* 276(1659): 1077–82.

Aragona, B. J., and Z. Wang (2009). "Dopamine regulation of social choice in a monogamous rodent species." *Front Behav Neurosci* 3: 15.

Araujo, A. B., V. Kupelian, et al. (2007). "Sex steroids and all-cause and cause-specific mortality in men." *Arch Intern Med* 167(12): 1252–60.

Araujo, A. B., T. G. Travison, et al. (2008). "Correlations between serum testosterone, estradiol, and sex-hormone-binding globulin and bone mineral density in a diverse sample of men." *J Clin Endocrinol Metab* 93(6): 2135–41.

Archer, J. (2009). "Does sexual selection explain human sex differences in aggression?" *Behav Brain Sci* 32(3–4): 249–66.

Archer, J. (2006). "Testosterone and human aggression: An evaluation of the challenge hypothesis." *Neurosci Biobehav Rev* 30(3): 319–45.

Arendash, G. W., and R. A. Gorski (1983). "Effects of discrete lesions of the sexually dimorphic nucleus of the preoptic area or other medial preoptic regions on the sexual behavior of male rats." *Brain Res Bull* 10(1): 147–54.

Arnold, A. P. (2009a). "The organizational-activational hypothesis as the foundation for a unified theory of sexual differentiation of all mammalian tissues." *Horm Behav* 55(5): 570–78; discussion 567–69.

Arnold, A. P. (2004). "Sex chromosomes and brain gender." *Nat Rev Neurosci* 5(9): 701–8.

Arnold, A. P., and X. Chen (2009b). "What does the 'four core genotypes' mouse model tell us about sex differences in the brain and other tissues?" *Front Neuroendocrinol* 30(1): 1–9.

Arnold, A. P., Susan E. Fahrbach (2009c). *Hormones, Brain and Behavior*, 2d ed. New York: Cambridge University Press.

Arnow, B. A., J. E. Desmond, et al. (2002). "Brain activation and sexual arousal in healthy, heterosexual males." *Brain* 125(Pt. 5): 1014–23.

Aron, A., H. Fisher, et al. (2005). "Reward, motivation, and emotion systems associated with early-stage intense romantic love." *J Neurophysiol* 94(1): 327–37.

Ashley, V., and D. Swick (2009). "Consequences of emotional stimuli: age differences on pure and mixed blocks of the emotional Stroop." *Behav Brain Funct* 5: 14.

Ashwin, C., E. Chapman, et al. (2006). "Impaired recognition of negative basic emotions in autism: a test of the amygdala theory." *Soc Neurosci* 1(3–4): 349–63.

Ashwin, C., P. Ricciardelli, et al. (2009). "Positive and negative gaze perception in autism spectrum conditions." *Soc Neurosci* 4(2): 153–64.

Atkins, D. C., D. H. Baucom, et al. (2001a). "Understanding infidelity: Correlates in a national random sample." *J Fam Psychol* 15(4): 735–49.

Atkins, D. C., S. Dimidjian, et al., eds. (2001b). *Why do people have affairs? Recent research and future directions about attributions for extramarital involvement.* New York: Cambridge University Press.

Atkins, D. C., J. Yi, et al. (2005). "Infidelity in couples seeking marital therapy." *J Fam Psychol* 19(3): 470–73.

Auger, A. P., D. P. Hexter, et al. (2001). "Sex difference in the phosphorylation of cAMP response element binding protein (CREB) in neonatal rat brain." *Brain Res* 890(1): 110–17.

Auslander, B. A., S. L. Rosenthal, et al. (2005). "Sexual development and behaviors of adolescents." *Pediatr Ann* 34(10): 785–93.

Auyeung, B., S. Baron-Cohen, et al. (2009a). "Fetal testosterone and autistic traits." *British Journal of Psychology* 100(1): 1–22.

Auyeung, B., S. Baron-Cohen, et al. (2009b). "Fetal testosterone predicts sexually differentiated childhood behavior in girls and in boys." *Psychological Science* 20(2): 144–48.

Auyeung, B., S. Baron-Cohen, et al. (2006). "Foetal testosterone and the child systemizing quotient." *Eur J Endocrinol* 155(suppl. 1): S123-S130.

Azurmendi, A., F. Braza, et al. (2006). "Aggression, dominance, and affiliation: Their relationships with androgen levels and intelligence in 5-year-old children." *Horm Behav* 50(1): 132–40.

Azurmendi, A., F. Braza, et al. (2005). "Cognitive abilities, androgen levels, and body mass index in 5-year-old children." *Horm Behav* 48(2): 187–95.

Bailey, J. M., M. P. Dunne, et al. (2000). "Genetic and environmental influences on sexual orientation and its correlates in an Australian twin sample." *Journal of Personality and Social Psychology* 78(3): 524–36.

Baillargeon, R. H., M. Zoccolillo, et al. (2007). "Gender differences in physical aggression: A prospective population-based survey of children before and after 2 years of age." *Dev Psychol* 43(1): 13–26.

Baker, J. R., M. G. Bemben, et al. (2006). "Effects of age on testosterone responses to resistance exercise and musculoskeletal variables in men." *J Strength Cond Res* 20(4): 874–81.

Baker, R., and M. A. Bellis (1995). Human Sperm Competition: Copulation, Masturbation, and Infidelity." London & New York: Chapman & Hall.

Bales, K. L., P. M. Plotsky, et al. (2007). "Neonatal oxytocin manipulations have long-lasting, sexually dimorphic effects on vasopressin receptors." *Neuroscience* 144(1): 38–45.

Balthazart, J., C. A. Cornil, et al. (2009). "Estradiol, a key endocrine signal in the sexual differentiation and activation of reproductive behavior in quail." *J Exp Zool Part A Ecol Genet Physiol* 311(5): 323–45.

Bancroft, J. (2005). "The endocrinology of sexual arousal." *J Endocrinol* 186(3): 411–27.

Barclay, P., and R. Willer (2007). "Partner choice creates competitive altruism in humans." *Proc Biol Sci* 274(1610): 749–53.

Baron, N. S. (2004). "See you online: Gender issues in college student use of instant messaging." *Journal of Language and Social Psychology Special Issue: Language and Communication Technology* 23(4): 397–423.

Baron-Cohen, S. (2002). "The extreme male brain theory of autism." *Trends Cogn Sci* 6(6): 248–54.

Baron-Cohen, S., B. Auyeung, et al. (2009). "Fetal testosterone and autistic traits: A response to three fascinating commentaries." *Br J Psychol* 100(Pt. 1): 39–47.

Baron-Cohen, S., and A. Klin (2006). "What's so special about Asperger syndrome?" *Brain Cogn* 61(1): 1–4.

Baron-Cohen, S., R. C. Knickmeyer, et al. (2005). "Sex differences in the brain: Implications for explaining autism." *Science* 310(5749): 819–23.

Baron-Cohen, S., S. Lutchmaya, and R. Knickmeyer (2004a). *Prenatal testosterone in mind.* Cambridge, MA: MIT Press.

Baron-Cohen, S., S. Lutchmaya, et al. (2004b). *Prenatal testosterone in mind: Amniotic fluid studies.* Cambridge, MA: MIT Press.

Baron-Cohen, S., J. Richler, et al. (2003). "The systemizing quotient: An investigation of adults with Asperger syndrome or high-functioning autism, and normal sex differences." *Philos Trans R Soc Lond B Biol Sci* 358(1430): 361–74.

Baron-Cohen, S., and S. Wheelwright (2004c). "The empathy quotient: An investigation of adults with Asperger syndrome or high functioning autism, and normal sex differences." *J Autism Dev Disord* 34(2): 163–75.

Barraza, J. A., and P. J. Zak (2009). "Empathy toward strangers triggers oxytocin release and subsequent generosity." *Ann N Y Acad Sci* 1167: 182–89.

Barron, E., P. B. Yang, et al. (2009). "Adolescent and adult male spontaneous hyperactive rats (SHR) respond differently to acute and chronic methylphenidate (Ritalin)." *Int J Neurosci* 119(1): 40–58.

Bartels, A., and S. Zeki (2004). "The neural correlates of maternal and romantic love." *Neuroimage* 21(3): 1155–66.

Bartels, A., and S. Zeki (2000). "The neural basis of romantic love." *Neuroreport* 11(17): 3829–34.

Baskerville, T. A., and A. J. Douglas (2008). "Interactions between dopamine and oxytocin in the control of sexual behaviour." *Prog Brain Res* 170: 277–90.

Basson, R. (2005). "Women's sexual dysfunction: Revised and expanded definitions." *Cmaj* 172(10): 1327–33.

Bastiaansen, J. A., M. Thioux, et al. (2009). "Evidence for mirror systems in emotions." *Philos Trans R Soc Lond B Biol Sci* 364(1528): 2391–2404.

Bateson, M., and S. D. Healy (2005). "Comparative evaluation and its implications for mate choice." *Trends Ecol Evol* 20(12): 659–64.

Baumeister, R. F., and D. Dhavale, eds. (2001). *Two sides of Romantic Rejection.* New York: Oxford University Press.

Bayliss, A. P., G. di Pellegrino, et al. (2005). "Sex differences in eye gaze and symbolic cueing of attention." *Q J Exp Psychol A* 58(4): 631–50.

Beach, F. A. (1971). "Hormonal factors controlling the differentiation, development, and display of copulatory behavior in the ramstergig and related species. In: E. Tobach, L. R. Aronson, and E. Shaw, eds., *The Biopsychology of Development.* New York: Academic Press, pp. 249–96.

Beach, F. A. (1967). "Cerebral and hormonal control of reflexive mechanisms involved in copulatory behavior." *Physiol Rev* 47(2): 289–316.

Beaulieu-Prevost, D., and A. Zadra (2007). "Absorption, psychological boundaries and attitude towards dreams as correlates of dream recall: Two decades of research seen through a meta-analysis." *J Sleep Res* 16(1): 51–59.

Beauregard, M., J. Courtemanche, et al. (2009). "The neural basis of unconditional love." *Psychiatry Res* 172(2): 93–98.

Becker, J. B. (2009). "Sexual differentiation of motivation: A novel mechanism?" *Horm Behav* 55(5): 646–54.

Becker, J. B. (2008a). "Sex differences in motivation." In: J.B. Becker, K. Berkley, N. Geary, E. Hampson, J. P. Herman, and E. A. Young, eds., *Sex Differences in the Brain: From Genes to Behavior.* Oxford, UK: Oxford University Press, pp. 177–99.

Becker, J. B., K. Berkley, N. Geary, E. Hampson, J. P. Herman, and E. A. Young, eds. (2008b). *Sex Differences in the Brain: From Genes to Behavior.* Oxford, UK: Oxford University Press.

Beckman, M. (2004). "Neuroscience: Crime, culpability, and the adolescent brain." *Science* 305(5684): 596–99.

Beech, J. R., and M. W. Beauvois (2006). "Early experience of sex hormones as a predictor of reading, phonology, and auditory perception." *Brain and Language* 96(1): 49–58.

Behar, D. M., R. Villems, et al. (2008). "The dawn of human matrilineal diversity." *Am J Hum Genet* 82(5): 1130–40.

Behrens, T. E., L. T. Hunt, et al. (2009). "The computation of social behavior." *Science* 324(5931): 1160–64.

Behrens, T. E., L. T. Hunt, et al. (2008). "Associative learning of social value." *Nature* 456(7219): 245–49.

Belgorosky, A., and M. A. Rivarola (1987). "Changes in serum sex hormone-binding globulin and in serum non-sex hormone-binding globulin-bound testosterone during prepuberty in boys." *J Steroid Biochem* 27(1–3): 291–95.

Bell, E. C., M. C. Willson, et al. (2006). "Males and females differ in brain activation during cognitive tasks." *Neuroimage* 30(2): 529–38.

Belsky, J. (1981). "Early human experience: A family perspective." *Developmental Psychology* 17(1): 3–23.

Benenson, J. F., and A. Christakos (2003). "The greater fragility of females' versus males' closest same-sex friendships." *Child Dev* 74(4): 1123–29.

Benenson, J. F., H. Markovits, et al. (2009a). "Strength determines coalitional strategies in humans." *Proc Biol Sci* 276(1667): 2589–95.

Benenson, J. F., H. Markovits, et al. (2009b). "Males' greater tolerance of same-sex peers." *Psychol Sci* 20(2): 184–90.

Bengtsson, S., H. Berglund, et al. (2001). "Brain activation during odor perception in males and females." *Neuroreport* 12(9): 2027–33.

Bensafi, M., W. M. Brown, et al. (2003). "Sex-steroid derived compounds induce sex-specific effects on autonomic nervous system function in humans." *Behav Neurosci* 117(6): 1125–34.

Berchtold, N. C., D. H. Cribbs, et al. (2008). "Gene expression changes in the course of normal brain aging are sexually dimorphic." *Proc Natl Acad Sci U S A* 105(40): 15605–610.

Bereczkei, T., P. Gyuris, et al. (2004). "Sexual imprinting in human mate choice." *Proc Biol Sci* 271(1544): 1129–34.

Berenbaum, S. A., J. B. Becker, K. Berkley, N. Geary, E. Hampson, J. P. Herman, and E. A. Young (2008). "Sex differences in children's play." In: J. B. Becker, K. Berkley, N. Geary, E. Hampson, J. P. Herman, and E. A. Young, eds., *Sex Differences in the Brain: From Genes to Behavior*. Oxford, UK: Oxford University Press.

Berenbaum, S. A., and M. Hines (1992). "Early androgens are related to childhood sex-typed toy preferences." *Psychol Sci* 3: 203–6.

Berg, S. J., and K. E. Wynne-Edwards (2002). "Salivary hormone concentrations in mothers and fathers becoming parents are not correlated." *Horm Behav* 42(4): 424–36.

Berg, S. J., and K. E. Wynne-Edwards (2001). "Changes in testosterone, cortisol, and estradiol levels in men becoming fathers." *Mayo Clin Proc* 76(6): 582–92.

Berglund, H., P. Lindstrom, et al. (2008). "Male-to-female transsexuals show sex-atypical hypothalamus activation when smelling odorous steroids." *Cereb Cortex* 18(8): 1900–1908.

Berglund, H., P. Lindstrom, et al. (2006). "Brain response to putative pheromones in lesbian women." *Proc Natl Acad Sci U S A* 103(21): 8269–74.

Berman, M., B. Gladue, et al. (1993). "The effects of hormones, type A behavior pattern, and provocation on aggression in men." *Motivation and Emotion* 17(2): 125–38.

Bernhardt, E. M., and F. K. Goldscheider (2001). "Men, resources, and family living: The determinants of union and parental status in the United States and Sweden." *Journal of Marriage & the Family* 63(3): 793–803.

Bernhardt, E. M., F. K. Goldscheider, et al. (2002). "Qualities men prefer for children in the US and Sweden: Differences among biological, step and informal fathers." *Journal of Comparative Family Studies* 33(2): 233–47.

Bernhardt, P. C. (1997). "Influences of serotonin and testosterone in aggression and dominance: Convergence with social psychology." *Current Directions in Psychological Science* 6(2): 44–48.

Bernhardt, P. C., J. M. Dabbs Jr., et al. (1998). "Testosterone changes during vicarious experiences of winning and losing among fans at sporting events." *Physiology & Behavior* 65(1): 59–62.

Berns, G. S., S. Moore, et al. (2009). "Adolescent engagement in dangerous behaviors is associated with increased white matter maturity of frontal cortex." *PLoS One* 4(8): e6773.

Berridge, K. C., and M. L. Kringelbach (2008). "Affective neuroscience of pleasure: Reward in humans and animals." *Psychopharmacology (Berl)* 199(3): 457–80.

Bertolino, A., G. Arciero, et al. (2005). "Variation of human amygdala response during threatening stimuli as a function of 5'HTTLPR genotype and personality style." *Biol Psychiatry* 57(12): 1517–25.

Bester-Meredith, J. K., and C. A. Marler (2003). "Vasopressin and the transmission of paternal behavior across generations in mated, cross-fostered *Peromyscus* mice." *Behav Neurosci* 117(3): 455–63.

Bester-Meredith, J. K., and C. A. Marler (2001). "Vasopressin and aggression in cross-fostered California mice (*Peromyscus californicus*) and white-footed mice (*Peromyscus leucopus*)." *Horm Behav* 40(1): 51–64.

Bianchi-Demicheli, F., and S. Ortigue (2007). "Toward an understanding of the cerebral substrates of woman's orgasm." *Neuropsychologia* 45(12): 2645–59.

Bingham, B., and V. Viau (2008). "Neonatal gonadectomy and adult testosterone replacement suggest an involvement of limbic arginine vasopressin and androgen receptors in the organization of the hypothalamic-pituitary-adrenal axis." *Endocrinology* 149(7): 3581–91.

Birmingham, W., B. N. Uchino, et al. (2009). "Social ties and cardiovascular function: An examination of relationship positivity and negativity during stress." *Int J Psychophysiol* 74(2): 114–19.

Bjorkqvist, K. (2001). "Social defeat as a stressor in humans." *Physiol Behav* 73(3): 435–42.

Bjorkqvist, K., M. Lindstrom, et al. (2000). "Attribution of aggression to acts: A four-factor model." *Psychol Rep* 87(2): 525–30.

Blanchard, R., and R. A. Lippa (2008). "The sex ratio of older siblings in non-right-handed homosexual men." *Arch Sex Behav* 37(6): 970–76.

Blanton, R. E., J. G. Levitt, et al. (2004). "Gender differences in the left inferior frontal gyrus in normal children." *Neuroimage* 22(2): 626–36.

Bleay, C., T. Comendant, et al. (2007). "An experimental test of frequency-dependent selection on male mating strategy in the field." *Proc Biol Sci* 274(1621): 2019–25.

Bocklandt, S., S. Horvath, et al. (2006). "Extreme skewing of X chromosome inactivation in mothers of homosexual men." *Hum Genet* 118(6): 691–94.

Bocklandt, S., and E. Vilain (2007). "Sex differences in brain and behavior: Hormones versus genes." *Adv Genet* 59: 245–66.

Bolona, E. R., M. V. Uraga, et al. (2007). "Testosterone use in men with sexual dysfunction: A systematic review and meta-analysis of randomized placebo-controlled trials." *Mayo Clin Proc* 82(1): 20–28.

Bolshakov, V. Y. (2009). "Nipping fear in the bud: Inhibitory control in the amygdala." *Neuron* 61(6): 817–19.

Boomsma, D. I., G. Willemsen, et al. (2005). "Genetic and environmental contributions to loneliness in adults: The Netherlands twin register study." *Behav Genet* 35(6): 745–52.

Boothroyd, L. G., B. C. Jones, D. M. Burt, and D. I. Perrett (2007). "Partner characteristics associated with masculinity, health and maturity in male faces." *Personality and Individual Differences* 43(5): 1161–73.

Borelli, J. L., and M. J. Prinstein (2006). "Reciprocal, longitudinal associations among adolescents' negative feedback-seeking, depressive symptoms, and peer relations." *J Abnorm Child Psychol* 34(2): 159–69.

Bornovalova, M. A., A. Cashman-Rolls, et al. (2009). "Risk taking differences on a behavioral task as a function of potential reward/loss magnitude and individual differences in impulsivity and sensation seeking." *Pharmacol Biochem Behav* 93(3): 258–62.

Bos, P. A., E. J. Hermans, et al. (2010). "Testosterone administration modulates neural responses to crying infants in young females." *Psychoneuroendocrinology* 35(1): 114–21.

Botwin, M. D., D. M. Buss, et al. (1997). "Personality and mate preferences: Five factors in mate selection and marital satisfaction." *J Pers* 65(1): 107–36.

Boulton, M., and R. Fitzpatrick, eds. (1996). *Bisexual Men in Britain.* Philadelphia: Taylor & Francis.

Boulton, M. J. (1996a). "Bullying in mixed sex groups of children." *Educational Psychology* 16(4): 439–43.

Boulton, M. J. (1996b). "A comparison of 8- and 11-year-old girls' and boys' participation in specific types of rough-and-tumble play and aggressive fighting: Implications for functional hypotheses." *Aggressive Behavior* 22(4): 271–87.

Bouma, E. M., H. Riese, et al. (2009). "Adolescents' cortisol responses to awakening and social stress; Effects of gender, menstrual phase and oral contraceptives: The TRAILS study." *Psychoneuroendocrinology* 34(6): 884–93.

Bouvattier, C., B. Mignot, et al. (2006). "Impaired sexual activity in male adults with partial androgen insensitivity." *J Clin Endocrinol Metab* 91(9): 3310–15.

Bouvattier, C., M. Tauber, et al. (1999). "Gonadotropin treatment of hypogonadotropic hypogonadal adolescents." *J Pediatr Endocrinol Metab* 12 Suppl 1: 339–44.

Bowlby, J. (1980). *Attachment and Loss.* New York: Basic Books.

Boyce, P., J. Condon, et al. (2007). "First-time fathers' study: Psychological distress in expectant fathers during pregnancy." *Aust N Z J Psychiatry* 41(9): 718–25.

Brambilla, D. J., A. M. Matsumoto, et al. (2009). "The effect of diurnal variation on clinical measurement of serum testosterone and other sex hormone levels in men." *J Clin Endocrinol Metab* 94(3): 907–13.

Bredy, T. W., R. E. Brown, et al. (2007). "Effect of resource availability on biparental care, and offspring neural and behavioral development in the California mouse (*Peromyscus californicus*)." *Eur J Neurosci* 25(2): 567–75.

Breedlove, S. M., and A. P. Arnold (1983). "Hormonal control of a developing neuromuscular system pt. 2: Sensitive periods for the androgen-induced masculinization of the rat spinal nucleus of the bulbocavernosus." *J Neurosci* 3(2): 424–32.

Breedlove, S. M., and A. P. Arnold (1980). "Hormone accumulation in a sexually dimorphic motor nucleus of the rat spinal cord." *Science* 210(4469): 564–66.

Brennan, P. A., and E. B. Keverne (2004). "Something in the air? New insights into mammalian pheromones." *Curr Biol* 14(2): R81–89.

Brenner, M., and D. R. Omark (1979). "The effects of sex, structure and social interaction on preschoolers' play behaviors in a naturalistic setting." *Instructional Science* 8(1): 91–105.

Bretherton, I., J. D. Lambert, et al. (2005). "Involved fathers of preschool children as seen by themselves and their wives: Accounts of attachment, socialization, and companionship." *Attach Hum Dev* 7(3): 229–51.

Bridges, R. S. (2008). "The effects of paternal behavior on offspring aggression and hormones in the biparental California mouse." In: Marler, C.A., B. C. Trainor, E. D. Gleason, J. K. Bester-Meredith, and E. A. Becker, eds., *Neurobiology of the Parental Brain*, Elsevier, ch. 28, pp. 435–48.

Briton, N. J., and J. A. Hall (1995). "Beliefs about female and male nonverbal communication." *Sex Roles* 32: 79–90.

Broad, K. D., J. P. Curley, et al. (2006). "Mother-infant bonding and the evolution of mammalian social relationships." *Philos Trans R Soc Lond B Biol Sci* 361(1476): 2199–2214.

Broad, K. D., and E. B. Keverne (2008). "More to pheromones than meets the nose." *Nat Neurosci* 11(2): 128–29.

Broaders, S. C., S. W. Cook, et al. (2007). "Making children gesture brings out implicit knowledge and leads to learning." *Journal of Experimental Psychology: General* 136(4): 539–50.

Brod, H., ed. (1997). *Pornography and the Alienation of Male Sexuality.* New York: New York University Press.

Brod, H., ed. (1987a). *A Case for Men's Studies.* Thousand Oaks, CA: Sage Publications.

Brod, H. ed. (1987b). *The Making of Masculinities: The New Men's Studies.* Boston: Allen & Unwin.

Brody, L., and J. A. Hall (1993). "Gender and emotion." In: M. Lewis and J. Haviland, eds., *Handbook of Emotions.* New York: Guilford, pp. 447–60.

Brody, S., and R. M. Costa (2009). "Satisfaction (sexual, life, relationship, and mental health) is associated directly with penile-vaginal intercourse, but inversely with other sexual behavior frequencies." *J Sex Med* 6(7): 1947–54.

Brooks, A., B. Schouten, et al. (2008). "Correlated changes in perceptions of the gender and orientation of ambiguous biological motion figures." *Curr Biol* 18(17): R728–R729.

Browning, J. R., E. Hatfield, et al. (2000). "Sexual motives, gender, and sexual behavior." *Archives of Sexual Behavior* 29(2): 135–53.

Buist, A., C. A. Morse, et al. (2003). "Men's adjustment to fatherhood: Implications for obstetric health care." *J Obstet Gynecol Neonatal Nurs* 32(2): 172–80.

Burch, R. L., and G. G. Gallup Jr. (2008). *Semen Science.* Hamburg, Germany: Springer.

Burch, R. L., and G. G. Gallup Jr., eds. (2006). *The Psychobiology of Human Semen.* New York: Cambridge University Press.

Burgdorf, J., and J. Panksepp (2001). "Tickling induces reward in adolescent rats." *Physiol Behav* 72(1–2): 167–73.

Burnett, S., G. Bird, et al. (2009a). "Development during adolescence of the neural processing of social emotion." *J Cogn Neurosci* 21(9): 1736–50.

Burnett, S., and S. J. Blakemore (2009b). "Functional connectivity during a social emotion task in adolescents and in adults." *Eur J Neurosci* 29(6): 1294–301.

Burnett, S., and S. J. Blakemore (2009c). "The development of adolescent social cognition." *Ann N Y Acad Sci* 1167: 51–56.

Burnham, T. C., J. F. Chapman, et al. (2003). "Men in committed, romantic relationships have lower testosterone." *Horm Behav* 44(2): 119–22.

Burri, A., M. Heinrichs, et al. (2008). "The acute effects of intranasal oxytocin administration on endocrine and sexual function in males." *Psychoneuroendocrinology* 33(5): 591–600.

Burriss, R. P., and A. C. Little (2006). "Effects of partner conception risk phase on male perception of dominance in faces." *Evolution and Human Behavior* 27(4): 297–305.

Buss, C., C. Lord, et al. (2007). "Maternal care modulates the relationship between prenatal risk and hippocampal volume in women but not in men." *J Neurosci* 27(10): 2592–95.

Buss, D. (1990). "International preferences in selecting mates: A study of 37 cultures." *Journal of Cross-Cultural Psychology* 21: 5–47.

Buss, D. M., ed. (2005a). *The Handbook of Evolutionary Psychology.* Hoboken, NJ: John Wiley.

Buss, D. M., ed. (2005b). *The Strategies of Human Mating.* Sunderland, MA: Sinauer Associates.

Buss, D. M. (2002). "Review of *Human Mate Guarding.*" *Neuro Endocrinol Lett* 23 Suppl. 4: 23–29.

Buss, D. M. (1995). "Psychological sex differences: Origins through sexual selection." *Am Psychol* 50(3): 164–68; discussion 169–71.

Buss, D. M. (1989). "Conflict between the sexes: strategic interference and the evocation of anger and upset." *J Pers Soc Psychol* 56(5): 735–47.

Buss, D. M., and M. Haselton (2005). *The Evolution of Jealousy: Comment.* Amsterdam: Elsevier Science.

Buss, D. M., and D. P. Schmitt (1993). "Sexual Strategies Theory: An evolutionary perspective on human mating." *Psychological Review* 100(2): 204–32.

Buss, D. M., T. K. Shackelford, et al. (2008). "The mate retention inventory–short form (MRI-SF)." *Personality and Individual Differences* 44(1): 322–34.

Byers, E. S., and S. MacNeil (2006). "Further validation of the interpersonal exchange model of sexual satisfaction." *J Sex Marital Ther* 32(1): 53–69.

Byrd-Craven, J., and D. C. Geary (2007). "Biological and evolutionary contributions to developmental sex differences." *Reprod Biomed Online* 15, Suppl. 2: 12–22.

Cabrera, N. J., C. S. Tamis-LeMonda, et al. (2000). "Fatherhood in the twenty-first century." *Child Dev* 71(1): 127–36.

Cacioppo, J. T., and J. Decety (2009a). "What are the brain mechanisms on which psychological processes are based?" *Perspectives on Psychological Science* 4(1): 10–18.

Cacioppo, J. T., and L. C. Hawkley (2009b). "Perceived social isolation and cognition." *Trends Cogn Sci* 13(10): 447–54.

Cacioppo, J. T., C. J. Norris, et al. (2009c). "In the eye of the beholder: Individual differences in perceived social isolation predict regional brain activation to social stimuli." *J Cogn Neurosci* 21(1): 83–92.

Cahill, L. (2006). "Why sex matters for neuroscience." *Nat Rev Neurosci* 7(6): 477–84.

Cahill, L. (2005). "His brain, her brain." *Sci Am* 292(5): 40–47.

Cahill, L. (2003). "Sex-related influences on the neurobiology of emotionally influenced memory." *Ann N Y Acad Sci* 985: 163–73.

Cahill, L., L. Gorski, et al. (2004). "The influence of sex versus sex-related traits on long-term memory for gist and detail from an emotional story." *Conscious Cogn* 13(2): 391–400.

Cahill, L., M. Uncapher, et al. (2004). "Sex-related hemispheric lateralization of amygdala function in emotionally influenced memory: An FMRI investigation." *Learn Mem* 11(3): 261–66.

Calasanti, T., and N. King (2007). "'Beware of the estrogen assault': Ideals of old manhood in anti-aging advertisements." *Journal of Aging Studies* 21(4): 357–68.

Caldwell, H. K., H. J. Lee, et al. (2008). "Vasopressin: behavioral roles of an 'original' neuropeptide." *Prog Neurobiol* 84(1): 1–24.

Calogero, R. M., and J. K. Thompson (2009). "Potential implications of the objectification of women's bodies for women's sexual satisfaction." *Body Image* 6(2): 145–48.

Cameron, N. M., F. A. Champagne, et al. (2005). "The programming of individual differences in defensive responses and reproductive strategies in the rat through variations in maternal care." *Neurosci Biobehav Rev* 29(4–5): 843–65.

Campbell, A. (2006). "Sex differences in direct aggression: What are the psychological mediators?" *Aggression and Violent Behavior* 11(3): 237–64.

Campbell, A. (1995). "A few good men: Evolutionary psychology and female adolescent aggression." *Ethology and Sociobiology* 16: 99–123.

Campbell, B. C., H. Prossinger, et al. (2005). "Timing of pubertal maturation and the onset of sexual behavior among Zimbabwe school boys." *Arch Sex Behav* 34(5): 505–16.

Canli, T., J. E. Desmond, et al. (2002). "Sex differences in the neural basis of emotional memories." *Proc Natl Acad Sci U S A* 99(16): 10789–94.

Cannon, E. A., S. J. Schoppe-Sullivan, et al. (2008). "Parent characteristics as antecedents of maternal gatekeeping and fathering behavior." *Fam Process* 47(4): 501–19.

Cannon, M. (2009). "Contrasting effects of maternal and paternal age on offspring intelligence." *PLoS Med* 6(3): e42.

Cant, M. A., and R. A. Johnstone (2008). "Reproductive conflict and the separation of reproductive generations in humans." *Proc Natl Acad Sci U S A* 105(14): 5332–36.

Card, N. A., B. D. Stucky, et al. (2008). "Direct and indirect aggression during childhood and adolescence: A meta-analytic review of gender differences, intercorrelations, and relations to maladjustment." *Child Dev* 79(5): 1185–1229.

Carere, C., G. F. Ball, et al. (2007). "Sex differences in projections from preoptic area aromatase cells to the periaqueductal gray in Japanese quail." *J Comp Neurol* 500(5): 894–907.

Carlier, J. G., and O. P. Steeno (1985). "Oigarche: The age at first ejaculation." *Andrologia* 17(1): 104–6.

Carlson, A. A., M. B. Manser, et al. (2006). "Cortisol levels are positively associated with pup-feeding rates in male meerkats." *Proc Biol Sci* 273(1586): 571–77.

Carlson, A. A., A. F. Russell, et al. (2006). "Elevated prolactin levels immediately precede decisions to babysit by male meerkat helpers." *Horm Behav* 50(1): 94–100.

Carpenter, D., E. Janssen, et al. (2008). "Women's scores on the sexual inhibition/sexual excitation scales (SIS/SES): Gender similarities and differences." *J Sex Res* 45(1): 36–48.

Carre, J. M., and C. M. McCormick (2008). "Aggressive behavior and change in salivary testosterone concentrations predict willingness to engage in a competitive task." *Horm Behav* 54(3): 403–9.

Carre, J. M., S. K. Putnam, et al. (2009). "Testosterone responses to competition predict future aggressive behaviour at a cost to reward in men." *Psychoneuroendocrinology* 34(4): 561–70.

Carrere, S., and J. M. Gottman (1999). "Predicting divorce among newlyweds from the first three minutes of a marital conflict discussion." *Family Process* 38(3): 293–301.

Carter, C. S. (2007). "Sex differences in oxytocin and vasopressin: Implications for autism spectrum disorders?" *Behav Brain Res* 176(1): 170–86.

Carter, C. S. (1998). "Neuroendocrine perspectives on social attachment and love." *Psychoneuroendocrinology* 23(8): 779–818.

Carter, C. S. (1992). "Oxytocin and sexual behavior." *Neurosci Biobehav Rev* 16(2): 131–44.

Carter, C. S., A. J. Grippo, et al. (2008). "Oxytocin, vasopressin and sociality." *Prog Brain Res* 170: 331–36.

Carter, C. S., J. Harris, and S. W. Porges (2009). "Neural and evolutionary perspectives on empathy." In: J. Decety and W. J. Ickes, eds., *Social Neuroscience of Empathy*. Cambridge, MA: MIT Press, pp. 169–82.

Casey, M. B., R. L. Nuttall, et al. (2001). "Spatial-mechanical reasoning skills versus mathematical self-confidence as mediators of gender differences on mathematics subtests using cross-national gender-based items." *Journal for Research in Mathematics Education* 32(1): 28–57.

Casey, M. B., R. L. Nuttall, et al. (1999). "Evidence in support of a model that predicts how biological and environmental factors interact to influence spatial skills." *Developmental Psychology* 35(5): 1237–47.

Casey, M. B., R. L. Nuttall, et al. (1997). "Mediators of gender differences in mathematics college entrance test scores: A comparison of spatial skills with internalized beliefs and anxieties." *Developmental Psychology* 33(4): 669–80.

Casey, M. B., R. Nuttall, et al. (1995). "The influence of spatial ability on gender differences in mathematics college entrance test scores across diverse samples." *Developmental Psychology* 31(4): 697–705.

Cassidy, J., (2001). "Gender differences among newborns on a transient otoacoustic emissions test for hearing." *Journal of Music Therapy* 37: 28–35.

Cauffman, E. (2004). "The adolescent brain: Excuse versus explanation; comments on part 4." *Ann N Y Acad Sci* 1021: 160–61.

Celichowski, J., and H. Drzymala (2006). "Differences between properties of male and female motor units in the rat medial gastrocnemius muscle." *J Physiol Pharmacol* 57(1): 83–93.

Chakrabarti, B., and S. Baron-Cohen (2006). "Empathizing: Neurocognitive developmental mechanisms and individual differences." *Prog Brain Res* 156: 403–17.

Chakrabarti, B., F. Dudbridge, et al. (2009). "Genes related to sex steroids, neural growth, and social-emotional behavior are associated with autistic traits, empathy, and Asperger syndrome." *Autism Res* 2(3): 157–77.

Champagne, F. A., J. P. Curley, et al. (2009). "Paternal influence on female behavior: The role of Peg3 in exploration, olfaction, and neuroendocrine regulation of maternal behavior of female mice." *Behav Neurosci* 123(3): 469–80.

Champagne, F. A., I. C. Weaver, et al. (2006). "Maternal care associated with methylation of the estrogen receptor-alpha1b promoter and estrogen receptor-alpha expression in the medial preoptic area of female offspring." *Endocrinology* 147(6): 2909–15.

Charles, S. T., and L. L. Carstensen (2008). "Unpleasant situations elicit different emotional responses in younger and older adults." *Psychol Aging* 2(3): 495–504.

Charlesworth, W. R., and C. Dzur (1987). "Gender comparisons of preschoolers' behavior and resource utilization in group problem solving." *Child Development* 58(1): 191–200.

Charlier, T. D., G. F. Ball, et al. (2008). "Rapid action on neuroplasticity precedes behavioral activation by testosterone." *Horm Behav* 54(4): 488–95.

Chen, X., W. Grisham, et al. (2009). "X chromosome number causes sex differences in gene expression in adult mouse striatum." *Eur J Neurosci* 29(4): 768–76.

Cheng, Y., K. H. Chou, et al. (2009). "Sex differences in the neuroanatomy of human mirror-neuron system: A voxel-based morphometric investigation." *Neuroscience* 158(2): 713–20.

Cheng, Y., J. Decety, et al. (2007). "Sex differences in spinal excitability during observation of bipedal locomotion." *Neuroreport* 18(9): 887–90.

Cheng, Y., A. N. Meltzoff, et al. (2007). "Motivation modulates the activity of the human mirror-neuron system." *Cereb Cortex* 17(8): 1979–86.

Cheng, Y. W., O. J. Tzeng, et al. (2006). "Gender differences in the human mirror system: A magnetoencephalography study." *Neuroreport* 17(11): 1115–19.

Cherney, I. (2008). "Mom, let me play more computer games: They improve my mental rotation skills." *Sex Roles* 59(11–12), Dec. 2008, ArtID 76.

Cherrier, M. M., A. M. Matsumoto, et al. (2007). "Characterization of verbal and spatial memory changes from moderate to supraphysiological increases in serum testosterone in healthy older men." *Psychoneuroendocrinology* 32(1): 72–79.

Chipman, K., E. Hampson, et al. (2002). "A sex difference in reliance on vision during manual sequencing tasks." *Neuropsychologia* 40(7): 910–16.

Cho, M. M., A. C. DeVries, et al. (1999). "The effects of oxytocin and vasopressin on partner preferences in male and female prairie voles (*Microtus ochrogaster*)." *Behav Neurosci* 113(5): 1071–79.

Choi, J. (2003). "Processes underlying sex differences in route learning strategies in children and adolescents." *Personality and Individual Differences* 34(7): 1153–66.

Chong, T. T., R. Cunnington, et al. (2008). "fMRI adaptation reveals mirror neurons in human inferior parietal cortex." *Curr Biol* 18(20): 1576–80.

Choudhury, S., S. J. Blakemore, et al. (2006). "Social cognitive development during adolescence." *Soc Cogn Affect Neurosci* 1(3): 165–74.

Christakou, A., R. Halari, et al. (2009). "Sex-dependent age modulation of frontostriatal and temporo-parietal activation during cognitive control." *Neuroimage* 48(1): 223–36.

Chura, L. R., M. V. Lombardo, et al. (2010). "Organizational effects of fetal testosterone on human corpus callosum size and asymmetry." *Psychoneuroendocrinology* 35(1): 122–32.

Cialdini, R. B., and M. R. Trost, eds. (1998a). *Social Influence: Social Norms, Conformity and Compliance.* New York: McGraw-Hill.

Cialdini, R. B., W. Wosinska, et al. (1998b). "When social role salience leads to social role rejection: Modest self-presentation among women and men in two cultures." *Personality and Social Psychology Bulletin* 24(5): 473–81.

Ciofi, P., O. C. Lapirot, et al. (2007). "An androgen-dependent sexual dimorphism visible at puberty in the rat hypothalamus." *Neuroscience* 146(2): 630–42.

Ciumas, C., A. Linden Hirschberg, et al. (2009). "High fetal testosterone and sexually dimorphic cerebral networks in females." *Cereb Cortex* 19(5): 1167–74.

Clark, M. M., and B. G. Galef Jr. (2000). "Why some male Mongolian gerbils may help at the nest: Testosterone, asexuality and alloparenting." *Anim Behav* 59(4): 801–6.

Clark, M. M., and B. G. Galef Jr. (1999). "A testosterone-mediated trade-off between parental and sexual effort in male Mongolian gerbils (*Meriones unguiculatus*)." *J Comp Psychol* 113(4): 388–95.

Clements-Stephens, A. M., S. L. Rimrodt, et al. (2009). "Developmental sex differences in basic visuospatial processing: Differences in strategy use?" *Neurosci Lett* 449(3): 155–60.

Clutton-Brock, T. H., S. J. Hodge, et al. (2006). "Intrasexual competition and sexual selection in cooperative mammals." *Nature* 444(7122): 1065–68.

Clutton-Brock, T. H., and K. Isvaran (2006). "Paternity loss in contrasting mammalian societies." *Biol Lett* 2(4): 513–16.

Coates, J. M., M. Gurnell, et al. (2009). "Second-to-fourth digit ratio predicts success among high-frequency financial traders." *Proc Natl Acad Sci U S A* 106(2): 623–28.

Coates, J. M., and J. Herbert (2008). "Endogenous steroids and financial risk taking on a London trading floor." *Proc Natl Acad Sci U S A* 105(16): 6167–72.

Cohen-Bendahan, C. C., J. K. Buitelaar, et al. (2005). "Is there an effect of prenatal testosterone on aggression and other behavioral traits? A study comparing same-sex and opposite-sex twin girls." *Horm Behav* 47(2): 230–37.

Cohen-Bendahan, C. C., J. K. Buitelaar, et al. (2004). "Prenatal exposure to testosterone and functional cerebral lateralization: A study in same-sex and opposite-sex twin girls." *Psychoneuroendocrinology* 29(7): 911–16.

Cohen-Bendahan, C. C., C. van de Beek, et al. (2005). "Prenatal sex hormone effects on child and adult sex-typed behavior: Methods and findings." *Neurosci Biobehav Rev* 29(2): 353–84.

Coiro, M. J., and R. E. Emery (1998). "Do marriage problems affect fathering more than mothering? A quantitative and qualitative review." *Clin Child Fam Psychol Rev* 1(1): 23–40.

Cole, W. R., S. H. Mostofsky, et al. (2008). "Age-related changes in motor subtle signs among girls and boys with ADHD." *Neurology* 71(19): 1514–20.

Collaer, M. L., and M. Hines (1995). "Human behavioral sex differences: A role for gonadal hormones during early development?" *Psychol Bull* 118(1): 55–107.

Collins, W. A., E. E. Maccoby, et al. (2001). *Toward Nature with Nurture.* New York: American Psychological Association.

Colson, M. H., A. Lemaire, et al. (2006). "Sexual behaviors and mental perception, satisfaction and expectations of sex life in men and women in France." *J Sex Med* 3(1): 121–31.

Condon, J. T., P. Boyce, et al. (2004). "The first-time fathers study: A prospective study of the mental health and wellbeing of men during the transition to parenthood." *Aust N Z J Psychiatry* 38(1–2): 56–64.

Connellan, J., S. Baron-Cohen, S (2000). "Sex differences in human neonatal social perception." *Infant Brain and Development* 23:113–18.

Conner, G. K., and V. Denson (1990). "Expectant fathers' response to pregnancy: Review of literature and implications for research in high-risk pregnancy." *J Perinat Neonatal Nurs* 4(2): 33–42.

Cook, S. W., and S. Goldin-Meadow (2006). "The role of gesture in learning: Do children use their hands to change their minds?" *Journal of Cognition and Development* 7(2): 211–32.

Cook, S. W., Z. Mitchell, et al. (2008). "Gesturing makes learning last." *Cognition* 106(2): 1047–58.

Cooke, B. (2005). "Sexually dimorphic synaptic organization of the medial amygdala." *Journal of Neuroscience* 25(46):10759–67.

Cornil, C. A., T. J. Stevenson, et al. (2009). "Are rapid changes in gonadal testosterone release involved in the fast modulation of brain estrogen effects?" *Gen Comp Endocrinol* 163(3): 298–305.

Corriveau, K., and P. L. Harris (2009). "Preschoolers continue to trust a more accurate informant 1 week after exposure to accuracy information." *Dev Sci* 12(1): 188–93.

Corty, E. W., and J. M. Guardiani (2008). "Canadian and American sex therapists' perceptions of normal and abnormal ejaculatory latencies: How long should intercourse last?" *J Sex Med* 5(5): 1251–56.

Cosgrove, K. P., C. M. Mazure, et al. (2007). "Evolving knowledge of sex differences in brain structure, function, and chemistry." *Biol Psychiatry* 62(8): 847–55.

Costa, R. M., and S. Brody (2009). "Greater frequency of penile-vaginal intercourse without condoms is associated with better mental health." *Arch Sex Behav*, published online July 28, 2009.

Cote, S. M., T. Vaillancourt, et al. (2006). "The development of physical aggression from toddlerhood to pre-adolescence: A nationwide longitudinal study of Canadian children." *J Abnorm Child Psychol* 34(1): 71–85.

Cousins, A. J., and S. W. Gangestad (2007). "Perceived threats of female infidelity, male proprietariness, and violence in college dating couples." *Violence and Victims* 22(6): 651–68.

Cox, D. L., and K. H. Bruckner (1999). *Women's Anger: Clinical and Developmental Perspectives*. Philadelphia: Brunner-Routledge.

Craig, H. K., and J. L. Evans (1991). "Turn exchange behaviors of children with normally developing language: The influence of gender." *J Speech Hear Res* 34(4): 866–78.

Craig, I. W., and K. E. Halton (2009). "Genetics of human aggressive behaviour." *Hum Genet* 126(1): 101–13.

Craig Roberts, S., A. C. Little, et al. (2009). "Manipulation of body odour alters men's self-confidence and judgements of their visual attractiveness by women." *Int J Cosmet Sci* 31(1): 47–54.

Crawford, J. (1992). *Emotion and Gender: Constructing Meaning from Memory*. London: Sage.

Crosby, R., R. Milhausen, et al. (2008). "Condom 'turn offs' among adults: An exploratory study." *Int J STD AIDS* 19(9): 590–94.

Crosby, R. A., W. L. Yarber, et al. (2007). "Men with broken condoms: Who and why?" *Sex Transm Infect* 83(1): 71–75.

Crowley, S. J., C. Acebo, et al. (2007). "Sleep, circadian rhythms, and delayed phase in adolescence." *Sleep Med* 8(6): 602–12.

Curtis, J. T., Y. Liu, et al. (2006). "Dopamine and monogamy." *Brain Res* 1126(1): 76–90.

Cushing, B. S., A. Perry, et al. (2008). "Estrogen receptors in the medial amygdala inhibit the expression of male prosocial behavior." *J Neurosci* 28(41): 10399–403.

Cushing, B. S., and K. E. Wynne-Edwards (2006). "Estrogen receptor-alpha distribution in male rodents is associated with social organization." *J Comp Neurol* 494(4): 595–605.

Dabbs Jr., J. M., M. F. Hargrove, et al. (1996). "Testosterone differences among college fraternities: Well-behaved vs rambunctious." *Personality and Individual Differences* 20(2): 157–61.

Dahl, R. E. (2008). "Biological, developmental, and neurobehavioral factors relevant to adolescent driving risks." *Am J Prev Med* 35(3, suppl.): S278–S284.

Dahl, R. E. (2004). "Adolescent brain development: a period of vulnerabilities and opportunities: Keynote address." *Ann N Y Acad Sci* 1021: 1–22.

Dalla, C., C. Edgecomb, et al. (2008). "Females do not express learned helplessness like males do." *Neuropsychopharmacology* 33(7): 1559–69.

Dalla, C., E. B. Papachristos, et al. (2009). "Female rats learn trace memories better than male rats and consequently retain a greater proportion of new neurons in their hippocampi." *Proc Natl Acad Sci U S A* 106(8): 2927–32.

Dalla, C., and T. J. Shors (2009). "Sex differences in learning processes of classical and operant conditioning." *Physiol Behav* 97(2): 229–38.

Dalla, C., A. S. Whetstone, et al. (2009). "Stressful experience has opposite effects on dendritic spines in the hippocampus of cycling versus masculinized females." *Neurosci Lett* 449(1): 52–56.

Danel, D., and B. Pawlowski (2007). "Eye-mouth-eye angle as a good indicator of face masculinization, asymmetry, and attractiveness (*Homo sapiens*)." *J Comp Psychol* 121(2): 221–25.

Danish, R. K., P. A. Lee, et al. (1980). "Micropenis, pt. 2: Hypogonadotropic hypogonadism." *Johns Hopkins Med J* 146(5): 177–84.

Dauwalder, B. (2008). "Systems behavior: Of male courtship, the nervous system and beyond in *Drosophila*." *Curr Genomics* 9(8): 517–24.

Dauwalder, B., S. Tsujimoto, et al. (2002). "The *Drosophila* takeout gene is regulated by the somatic sex-determination pathway and affects male courtship behavior." *Genes Dev* 16(22): 2879–92.

Davies, A.P.C., T. K. Shackelford, et al. (2007). "When a 'poach' is not a poach: Re-defining human mate poaching and re-estimating its frequency." *Archives of Sexual Behavior* 36(5): 702–16.

Davies, A.P.C., T. K. Shackelford, et al. (2006). " 'Attached' or 'unattached': With whom do men and women prefer to mate, and why?" *Psihologijske Teme* 15(2): 297–314.

Davis, J. A., and G. G. Gallup Jr., eds. (2006). *Preeclampsia and Other Pregnancy Complications as an Adaptive Response to Unfamiliar Semen.* New York: Cambridge University Press.

De Bellis, M. D., M. S. Keshavan, et al. (2001). "Sex differences in brain maturation during childhood and adolescence." *Cereb Cortex* 11(6): 552–57.

De Groot, B., and J. J. Duyvene De Wit (1949). "Copulin and ovipositor growth in the female bitterling (*Rhodeus amarus Bl*)." *Acta Endocrinol* (Copenh) 3(2): 129–36.

De Lisi, R., and J. L. Wolford (2002). "Improving children's mental rotation accuracy with computer game playing." *J Genet Psychol* 163(3): 272–82.

De Vries, G. J. (2005). "Sex steroids and sex chromosomes at odds?" *Endocrinology* 146(8): 3277–79.

De Vries, G. J. (2004). "Minireview: Sex differences in adult and developing brains—compensation, compensation, compensation." *Endocrinology* 145(3): 1063–68.

De Vries, G. J., and P. A. Boyle (1998). "Double duty for sex differences in the brain." *Behav Brain Res* 92(2): 205–13.

De Vries, G. J., M. Jardon, et al. (2008). "Sexual differentiation of vasopressin innervation of the brain: Cell death versus phenotypic differentiation." *Endocrinology* 149(9): 4632–37.

De Vries, G. J., and G. C. Panzica (2006). "Sexual differentiation of central vasopressin and vasotocin systems in vertebrates: Different mechanisms, similar endpoints." *Neuroscience* 138(3): 947–55.

De Vries, G. J., and P. Sodersten (2009). "Sex differences in the brain: The relation between structure and function." *Horm Behav* 55(5): 589–96.

De Zegher, F., H. Devlieger, et al. (1992). "Pulsatile and sexually dimorphic secretion of luteinizing hormone in the human infant on the day of birth." *Pediatr Res* 32(5): 605–7.

Debiec, J. (2005). "Peptides of love and fear: Vasopressin and oxytocin modulate the integration of information in the amygdala." *Bioessays* 27(9): 869–73.

DeCaro, D. A., M. Bar-Eli, et al. (2009). "How do motoric realities shape, and become shaped by, the way people evaluate and select potential courses of action? Toward a unitary framework of embodied decision making." *Prog Brain Res* 174: 189–203.

Decety, J., and C. Lamm (2007). "The role of the right temporoparietal junction in social interaction: How low-level computational processes contribute to meta-cognition." *Neuroscientist* 13(6): 580–93.

Decety, J., and C. Lamm (2006). "Human empathy through the lens of social neuroscience." *Scientific World Journal* 6: 1146–63.

Decety, J., and M. Meyer (2008). "From emotion resonance to empathic understanding: A social developmental neuroscience account." *Dev Psychopathol* 20(4): 1053–80.

Decety, J., K. J. Michalska, et al. (2009). "Atypical empathic responses in adolescents with aggressive conduct disorder: A functional MRI investigation." *Biol Psychol* 80(2): 203–11.

Dedovic, K., M. Wadiwalla, et al. (2009). "The role of sex and gender socialization in stress reactivity." *Developmental Psychology* 45(1): 45–55.

Deeley, Q., E. M. Daly, et al. (2008). "Changes in male brain responses to emotional faces from adolescence to middle age." *Neuroimage* 40(1): 389–97.

Dekker, A., and G. Schmidt (2002). "Patterns of masturbatory behaviour: Changes between the sixties and the nineties." *Journal of Psychology & Human Sexuality* 14(2–3): 35–48.

Delahunty, K. M., D. W. McKay, et al. (2007). "Prolactin responses to infant cues in men and women: Effects of parental experience and recent infant contact." *Horm Behav* 51(2): 213–20.

Delville, Y., K. M. Mansour, et al. (1996). "Testosterone facilitates aggression by modulating vasopressin receptors in the hypothalamus." *Physiol Behav* 60(1): 25–29.

Dewing, P., C. W. Chiang, et al. (2006). "Direct regulation of adult brain function by the male-specific factor SRY." *Curr Biol* 16(4): 415–20.

Dewing, P., T. Shi, et al. (2003). "Sexually dimorphic gene expression in mouse brain precedes gonadal differentiation." *Brain Res Mol Brain Res* 118(1–2): 82–90.

Dharia, S., and C. R. Parker, Jr. (2004). "Adrenal androgens and aging." *Semin Reprod Med* 22(4): 361–68.

Diamond, J. (1997). *Why Is Sex Fun?* New York: Basic Books.

Diamond, M. J. (2006). "Masculinity unraveled: The roots of male gender identity and the shifting of male ego ideals throughout life." *J Am Psychoanal Assoc* 54(4): 1099–1130.

Dillon, B. E., N. B. Chama, et al. (2008). "Penile size and penile enlargement surgery: A review." *Int J Impot Res* 20(6): 519–29.

Dimberg, U., and L. O. Lundquist (1990). "Gender differences in facial reactions to facial expressions." *Biol Psychol* 30(2): 151–59.

DiPietro, J. A. (1981). "Rough and tumble play: A function of gender." *Developmental Psychology* 17(1): 50–58.

DiRocco, D. P., and Z. Xia (2007). "Alpha males win again." *Nat Neurosci* 10(8): 938–40.

Ditzen, B., C. Hoppmann, et al. (2008). "Positive couple interactions and daily cortisol: On the stress-protecting role of intimacy." *Psychosom Med* 70(8): 883–89.

Ditzen, B., M. Schaer, et al. (2009). "Intranasal oxytocin increases positive communication and reduces cortisol levels during couple conflict." *Biol Psychiatry* 65(9): 728–31.

Domes, G., M. Heinrichs, et al. (2007a). "Oxytocin attenuates amygdala responses to emotional faces regardless of valence." *Biol Psychiatry* 62(10): 1187–90.

Domes, G., M. Heinrichs, et al. (2007b). "Oxytocin improves 'mind-reading' in humans." *Biol Psychiatry* 61(6): 731–33.

Donaldson, Z. R., F. A. Kondrashov, et al. (2008). "Evolution of a behavior-linked microsatellite-containing element in the 5' flanking region of the primate AVPR1A gene." *BMC Evol Biol* 8: 180.

Donaldson, Z. R., and L. J. Young (2008). "Oxytocin, vasopressin, and the neurogenetics of sociality." *Science* 322(5903): 900–904.

Doremus-Fitzwater, T. L., E. I. Varlinskaya, et al. (2010). "Motivational systems in adolescence: Possible implications for age differences in substance abuse and other risk-taking behaviors." *Brain Cogn* 72(1): 114–23.

Dreber, A., C. L. Apicella, et al. (2009). "The 7R polymorphism in the dopamine receptor gene (DRD4) is associated with financial risk taking in men." *Evolution and Human Behavior* 30(2): 85–92.

Dreher, J. C., A. Meyer-Lindenberg, et al. (2008). "Age-related changes in midbrain dopaminergic regulation of the human reward system." *Proc Natl Acad Sci USA* 105(39): 15106–11.

Driver, J. L., and J. M. Gottman (2004). "Daily marital interactions and positive affect during marital conflict among newlywed couples." *Family Process* 43(3): 301–14.

Dugger, B. N., J. A. Morris, et al. (2008). "Gonadal steroids regulate neural plasticity in the sexually dimorphic nucleus of the preoptic area of adult male and female rats." *Neuroendocrinology* 88(1): 17–24.

Dunbar, R. I. (2009). "The social brain hypothesis and its implications for social evolution." *Ann Hum Biol* 36(5): 562–72.

Dunbar, R. I. (2007a). "Male and female brain evolution is subject to contrasting selection pressures in primates." *BMC Biol* 5: 21.

Dunbar, R. I., and S. Shultz (2007b). "Evolution in the social brain." *Science* 317(5843): 1344–47.

Eaton, D. K., L. Kann, et al. (2008). "Youth risk behavior surveillance—United States, 2007." *MMWR Surveill Summ* 57(4): 1–131.

Eaton, W. O., and L. R. Enns (1986). "Sex differences in human motor activity level." *Psychological Bulletin* 100(1): 19–28.

Eckel, L. (2008). "Hormone-behavior relations." In: J. B. Becker, K. Berkley, N. Geary, E. Hampson, J. P. Herman, and E. A. Young, eds. *Sex Differences in the Brain: From Genes to Behavior.* Oxford, UK: Oxford University Press.

Edelman, M. S., and D. R. Omark (1973). "Dominance hierarchies in young children." *Social Science Information/Sur les sciences sociales,* 7(1): 103–10.

Edelstein, D., M. Sivanandy, et al. (2007). "The latest options and future agents for treating male hypogonadism." *Expert Opin Pharmacother* 8(17): 2991–3008.

Ehrlich, S. B., S. C. Levine, et al. (2006). "The importance of gesture in children's spatial reasoning." *Dev Psychol* 42(6): 1259–68.

Eibl-Eibesfeldt, I., ed. (1972). *Similarities and differences between cultures in expressive movements.* Oxford, England: Cambridge University Press.

Eisenberger, N. I., and M. D. Lieberman (2004). "Why rejection hurts: A common neural alarm system for physical and social pain." *Trends Cogn Sci* 8(7): 294–300.

Ekman, P. F., (1978). *Facial Action Coding System: A Technique for the Measurement of Facial Movement.* Palo Alto, CA: Consulting Psychologists Press.

Ellis, L., and M. A. Ames (1987). "Neurohormonal functioning and sexual orientation: A theory of homosexuality-heterosexuality." *Psychol Bull* 101(2): 233–58.

Eme, R. F. (2007). "Sex differences in child-onset, life-course-persistent conduct disorder. A review of biological influences." *Clin Psychol Rev* 27(5): 607–27.

Emery Thompson, M., J. H. Jones, et al. (2007). "Aging and fertility patterns in wild chimpanzees provide insights into the evolution of menopause." *Curr Biol* 17(24): 2150–56.

Erlandsson, K., A. Dsilna, et al. (2007). "Skin-to-skin care with the father after cesarean birth and its effect on newborn crying and prefeeding behavior." *Birth* 34(2): 105–14.

Everhart, D. E., H. A. Demaree, et al. (2006). "Perception of emotional prosody: Moving toward a model that incorporates sex-related differences." *Behav Cogn Neurosci Rev* 5(2): 92–102.

Evuarherhe, O., J. D. Leggett, et al. (2009). "Organizational role for pubertal androgens on adult hypothalamic-pituitary-adrenal sensitivity to testosterone in the male rat." *J Physiol* 587(pt. 12): 2977–85.

Exton, M. S., T. H. Kruger, et al. (2001a). "Coitus-induced orgasm stimulates prolactin secretion in healthy subjects." *Psychoneuroendocrinology* 26(3): 287–94.

Exton, M. S., T. H. Kruger, et al. (2001b). "Endocrine response to masturbation-induced orgasm in healthy men following a 3-week sexual abstinence." *World J Urol* 19(5): 377–82.

Fabes, R. A., C. L. Martin, et al. (2003a). "Early school competence: The roles of sex-segregated play and effortful control." *Dev Psychol* 39(5): 848–58.

Fabes, R. A., C. L. Martin, et al. (2003b). "Young children's play qualities in same-, other-, and mixed-sex peer groups." *Child Dev* 74(3): 921–32.

Fagan, J., R. Palkovitz, et al. (2009). "Pathways to paternal engagement: Longitudinal effects of risk and resilience on nonresident fathers." *Dev Psychol* 45(5): 1389–405.

Fagot, B. I., and R. Hagan (1985). "Aggression in toddlers: Responses to the assertive acts of boys and girls." *Sex Roles* 12(3–4): 341–51.

Farrow, T. F., R. Reilly, et al. (2003). "Sex and personality traits influence the difference between time taken to tell the truth or lie." *Percept Mot Skills* 97(2): 451–60.

Feinberg, D. R., L. M. DeBruine, et al. (2008). "The role of femininity and averageness of voice pitch in aesthetic judgments of women's voices." *Perception* 37(4): 615–23.

Feiring, C., and M. Lewis (1987). "The child's social network: Sex differences from three to six years." *Sex Roles* 17(11–12): 621–36.

Felder, S. (2006). "The gender longevity gap: explaining the difference between singles and couples." *Journal of Population Economics* 19(3): 1432–75.

Feldman, R. (2007). "Parent-infant synchrony and the construction of shared timing: Physiological precursors, developmental outcomes, and risk conditions." *J Child Psychol Psychiatry* 48(3–4): 329–54.

Feldman, R. (2006). "From biological rhythms to social rhythms: Physiological precursors of mother-infant synchrony." *Dev Psychol* 42(1): 175–88.

Feldman, R., and A. I. Eidelman (2009). "Biological and environmental initial conditions shape the trajectories of cognitive and social-emotional development across the first years of life." *Dev Sci* 12(1): 194–200.

Feldman, R., A. I. Eidelman, et al. (2002). "Comparison of skin-to-skin (kangaroo) and traditional care: Parenting outcomes and preterm infant development." *Pediatrics* 110(1, pt. 1): 16–26.

Feldman, R., A. Weller, et al. (2003). "Testing a family intervention hypothesis: The contribution of mother-infant skin-to-skin contact (kangaroo care) to family interaction, proximity, and touch." *J Fam Psychol* 17(1): 94–107.

Feng, J., I. Spence, et al. (2007). "Playing an action video game reduces gender differences in spatial cognition." *Psychol Sci* 18(10): 850–55.

Ferguson, R., and C. O'Neill, eds. (2001). *Late Adolescence: A Gestalt Model of Development, Crisis, and Brief Psychotherapy*. Cambridge, MA: Gestalt Press Book; New York: Analytic Press/Taylor & Francis.

Ferguson, T., and H. Eyre (2000). "Engendering gender differences in shame and guilt: Stereotypes, socialization and situational pressures." In: A. H. Fisher, ed., *Gender and Emotion: Social Psychological Perspectives*. Cambridge, UK: Cambridge University Press, 2000, pp. 254–76.

Fernald, A., T. Taeschner, et al. (1989). "A cross-language study of prosodic modifications in mothers' and fathers' speech to preverbal infants." *J Child Lang* 16(3): 477–501.

Fernandez-Guasti, A., F. P. Kruijver, et al. (2000). "Sex differences in the distribution of androgen receptors in the human hypothalamus." *J Comp Neurol* 425(3): 422–35.

Ferris, C. F. (2008a). "Functional magnetic resonance imaging and the neurobiology of vasopressin and oxytocin." *Prog Brain Res* 170: 305–20.

Ferris, C. F., Y. Delville, et al. (1996). "Vasopressin and developmental onset of flank marking behavior in golden hamsters." *J Neurobiol* 30(2): 192–204.

Ferris, C. F., C. T. Snowdon, et al. (2004). "Activation of neural pathways associated with sexual arousal in non-human primates." *J Magn Reson Imaging* 19(2): 168–75.

Ferris, C. F., T. Stolberg, et al. (2008b). "Imaging the neural circuitry and chemical control of aggressive motivation." *BMC Neurosci* 9: 111.

Feygin, D. L., J. E. Swain, et al. (2006). "The normalcy of neurosis: evolutionary origins of obsessive-compulsive disorder and related behaviors." *Prog Neuropsychopharmacol Biol Psychiatry* 30(5): 854–64.

Fischer, A. H., ed. (2000). *Gender and Emotion*. Paris: Cambridge University Press.

Field, E. F. (2008). "Sex differences in the organization of movement." In: J. B. Becker, K. Berkley, N. Geary, E. Hampson, J. P. Herman, and E. A. Young, eds., *Sex Differences in the Brain: From Genes to Behavior*, Oxford, UK: Oxford University Press.

Field, E. F., I. Q. Whishaw, et al. (1997). "A kinematic analysis of sex-typical movement patterns used during evasive dodging to protect a food item: The role of testicular hormones." *Behav Neurosci* 111(4): 808–15.

Finegan, J. A., B. Bartleman, et al. (1989). "A window for the study of prenatal sex hormone influences on postnatal development." *J Genet Psychol* 150(1): 101–12.

Finkelstein, J. W., E. J. Susman, et al. (1997). "Estrogen or testosterone increases self-reported aggressive behaviors in hypogonadal adolescents." *J Clin Endocrinol Metab* 82(8): 2433–38.

Finn, M., and K. Henwood (2009). "Exploring masculinities within men's identificatory imaginings of first-time fatherhood." *Br J Soc Psychol* 48(pt. 3): 547–62.

Fischer, A. R. (2007). "Parental relationship quality and masculine gender-role strain in young men: Mediating effects of personality." *Counseling Psychologist* 35(2): 328–58.

Fisher, H. (2006). Personal communication.

Fisher, H. (2004). *Why We Love: The Nature and Chemistry of Romantic Love.* New York: Holt.

Fisher, H., A. Aron, et al. (2005). "Romantic love: An fMRI study of a neural mechanism for mate choice." *J Comp Neurol* 493(1): 58–62.

Fisher, H. E., A. Aron, et al. (2002). "Defining the brain systems of lust, romantic attraction, and attachment." *Arch Sex Behav* 31(5): 413–19.

Fiske, S. T. (2009). "Brain scans of male brain: Area for empathy shut down after looking at sexy pictures . . . suggest sexy images can shift the way men perceive women." Paper presented at meeting of the American Association for the Advancement of Science, Chicago, Feb. 15, 2008.

Flanders, J. L., V. Leo, et al. (2009). "Rough-and-tumble play and the regulation of aggression: An observational study of father-child play dyads." *Aggress Behav* 35(4): 285–95.

Fleming, A. S., C. Corter, et al. (2002). "Testosterone and prolactin are associated with emotional responses to infant cries in new fathers." *Horm Behav* 42(4): 399–413.

Fleming, A. S., D. H. O'Day, et al. (1999). "Neurobiology of mother-infant interactions: Experience and central nervous system plasticity across development and generations." *Neurosci Biobehav Rev* 23(5): 673–85.

Fliers, E., G. J. De Vries, et al. (1985). "Changes with aging in the vasopressin and oxytocin innervation of the rat brain." *Brain Res* 348(1): 1–8.

Forger, N. G. (2009). "Control of cell number in the sexually dimorphic brain and spinal cord." *J Neuroendocrinol* 21(4): 393–99.

Forger, N. G. (2009). "The organizational hypothesis and final common pathways: Sexual differentiation of the spinal cord and peripheral nervous system." *Horm Behav* 55(5): 605–10.

Forger, N. G. (2006). "Cell death and sexual differentiation of the nervous system." *Neuroscience* 138(3): 929–38.

Fox, A. B., D. Bukatko, et al. (2007). "The medium makes a difference: Gender similarities and differences in instant messaging." *Journal of Language and Social Psychology* 26(4): 389–97.

Francken, A. B., H. B. van de Wiel, et al. (2002). "What importance do women attribute to the size of the penis?" *Eur Urol* 42(5): 426–31.

Frazier, C. R., B. C. Trainor, et al. (2006). "Paternal behavior influences development of aggression and vasopressin expression in male California mouse offspring." *Horm Behav* 50(5): 699–707.

Frederick, D. A., J. Lever, et al. (2007). "Interest in cosmetic surgery and body image: Views of men and women across the lifespan." *Plast Reconstr Surg* 120(5): 1407–15.

Frederick, D. A., L. A. Peplau, et al. (2006). "The swimsuit issue: Correlates of body image in a sample of 52,677 heterosexual adults." *Body Image* 3(4): 413–19.

Freeman, J. B., N. O. Rule, et al. (2009a). "Culture shapes a mesolimbic response to signals of dominance and subordination that associates with behavior." *Neuroimage* 47(1): 353–59.

Freeman, J. B., D. Schiller, et al. (2009b). "The neural origins of superficial and individuated judgments about ingroup and outgroup members." *Hum Brain Mapp* 31(1): 150–59.

Frey, K. A., S. M. Navarro, et al. (2008). "The clinical content of preconception care: Preconception care for men." *Am J Obstet Gynecol* 199(6, suppl. 2): S389–S395.

Frey, W. (1985). *Crying: The Mystery of Tears.* Minneapolis: Winston.

Frosh, S., A. Phoenix, et al. (2005). "Struggling towards manhood: Narratives of homophobia and fathering." *British Journal of Psychotherapy, Special Issue: Masculinity* 22(1): 37–55.

Fukushima, H., and K. Hiraki (2009). "Whose loss is it? Human electrophysiological correlates of non-self reward processing." *Soc Neurosci:* 1–15.

Fukushima, H., and K. Hiraki (2006). "Perceiving an opponent's loss: Gender-related differences in the medial-frontal negativity." *Soc Cogn Affect Neurosci* 1(2): 149–57.

Fuxjager, M. J., G. Mast, et al. (2009). "The 'home advantage' is necessary for a full winner effect and changes in post-encounter testosterone." *Horm Behav* 56(2): 214–19.

Fynn-Thompson, E., H. Cheng, et al. (2003). "Inhibition of steroidogenesis in Leydig cells by Müllerian-inhibiting substance." *Mol Cell Endocrinol* 211(1–2): 99–104.

Gabory, A., L. Attig, et al. (2009). "Sexual dimorphism in environmental epigenetic programming." *Mol Cell Endocrinol* 304(1–2): 8–18.

Gagnidze, K., and D. W. Pfaff (2009). "Sex on the brain." *Cell* 139(1): 19–21.

Gagnon, M. D., M. Hersen, et al. (1999). "Interpersonal and psychological correlates of marital dissatisfaction in late life: A review." *Clin Psychol Rev* 19(3): 359–78.

Gallup, G. G., Jr. (2008). "Kissing." In: H. T. Reis and S. Sprecher, eds., *Encyclopedia of Human Relations.* Thousand Oaks, CA: Sage Publications.

Gallup Jr., G. G., and R. L. Burch, eds. (2006a). *The Semen-Displacement Hypothesis: Semen Hydraulics and the Intra-Pair Copulation Proclivity Model of Female Infidelity.* New York: Cambridge University Press.

Gallup Jr., G. G., R. L. Burch, et al. (2006b). "Semen displacement as a sperm competition strategy: Multiple mating, self-semen displacement, and timing of in-pair copulations." *Human Nature, special issue: Human sperm competition* 17(3): 253–64.

Gallup Jr., G. G., and R. L. Burch (2004). "Semen displacement as a sperm competition strategy in humans." *Evolutionary Psychology* 2: 12–23.

Gallup Jr, G. G., R. L. Burch, et al. (2003). "The human penis as a semen displacement device." *Evolution and Human Behavior* 24(4): 277–89.

Gallup Jr., G. G., R. L. Burch, et al. (2002). "Does semen have antidepressant properties?" *Archives of Sexual Behavior* 31(3): 289–93.

Gangestad, S. W., ed. (2006). *Evidence for Adaptations for Female Extra-Pair Mating in Humans: Thoughts on Current Status and Future Directions.* New York: Cambridge University Press.

Gangestad, S. W. (2000). "Human sexual selection, good genes, and special design." *Ann NY Acad Sci* 907: 50–61.

Gangestad, S. W. (1993). "Sexual selection and physical attractiveness: Implications for mating dynamics." *Human Nature* 4(3): 205–35.

Gangestad, S. W., C. E. Garver-Apgar, et al. (2007). "Changes in women's mate preferences across the ovulatory cycle." *Journal of Personality and Social Psychology* 92(1): 151–63.

Gangestad, S. W., M. G. Haselton, et al. (2006). "Evolutionary foundations of cultural variation: Evoked culture and mate preferences." *Psychological Inquiry* 17(2): 75–95.

Gao, X., P. Phillips, et al. (1994). "Androgen manipulation and vasopressin binding in the rat brain and peripheral organs." *Eur J Endocrinol* 130(3): 291–96.

Garcia-Falgueras, A., and D. F. Swaab (2008). "A sex difference in the hypothalamic uncinate nucleus: Relationship to gender identity." *Brain* 131(Pt. 12): 3132–46.

Garrett, B. (2009). *Brain and Behavior: An Introduction to Biological Psychology*, 2d ed. Thousand Oaks, CA: Sage Publications.

Garver-Apgar, C. E., S. W. Gangestad, et al. (2008). "Hormonal correlates of women's mid-cycle preference for the scent of symmetry." *Evolution and Human Behavior* 29(4): 223–32.

Garver-Apgar, C. E., S. W. Gangestad, et al. (2007). "Women's perceptions of men's sexual coerciveness change across the menstrual cycle." *Acta Psychologica Sinica, special issue: Evolutionary psychology* 39(3): 536–40.

Garver-Apgar, C. E., S. W. Gangestad, et al. (2006). "Major histocompatibility complex alleles, sexual responsivity, and unfaithfulness in romantic couples." *Psychol Sci* 17(10): 830–35.

Gasbarri, A., B. Arnone, et al. (2007). "Sex-related hemispheric lateralization of electrical potentials evoked by arousing negative stimuli." *Brain Res* 1138C: 178–86.

Gasbarri, A., B. Arnone, et al. (2006). "Sex-related lateralized effect of emotional content on declarative memory: An event related potential study." *Behav Brain Res* 168(2): 177–84.

Gasbarri, A., A. Pompili, et al. (2008). "Working memory for emotional facial expressions: Role of the estrogen in young women." *Psychoneuroendocrinology* 33(7): 964–72.

Gatzke-Kopp, L. M., T. P. Beauchaine, et al. (2009). "Neurological correlates of reward responding in adolescents with and without externalizing behavior disorders." *J Abnorm Psychol* 118(1): 203–13.

Geary, D. (1998). *Male, female.* New York: APA Press.

Geary, D. C. (2000). "Evolution and proximate expression of human paternal investment." *Psychological Bulletin* 126(1): 55–77.

Geary, D. C., S. J. Saults, et al. (2000). "Sex differences in spatial cognition, computational fluency, and arithmetical reasoning." *Journal of Experimental Child Psychology, Special Issue: Sex and gender development* 77(4): 337–53.

Geary, D. C., J. Vigil, et al. (2004). "Evolution of human mate choice." *J Sex Res* 41(1): 27–42.

Geenen, V., F. Adam, et al. (1988). "Inhibitory influence of oxytocin infusion on contingent negative variation and some memory tasks in normal men." *Psychoneuroendocrinology* 13(5): 367–75.

Geiser, C., W. Lehmann, et al. (2008a). "A note on sex differences in mental rotation in different age groups." *Intelligence* 36(6): 556–63.

Geiser, C., W. Lehmann, et al. (2008b). "Quantitative and qualitative change in children's mental rotation performance." *Learning and Individual Differences* 18(4): 419–29.

Georgiadis, J. R., A. A. Reinders, et al. (2009). "Men versus women on sexual brain function: Prominent differences during tactile genital stimulation, but not during orgasm." *Hum Brain Mapp* 30(10): 3089–3101.

Gerressu, M., C. H. Mercer, et al. (2008a). "Prevalence of masturbation and associated factors in a British national probability survey." *Arch Sex Behav* 37(2): 266–78.

Gerressu, M. and J. M. Stephenson (2008b). "Sexual behaviour in young people." *Curr Opin Infect Dis* 21(1): 37–41.

Gesquiere, L. R., J. Altmann, et al. (2005). "Coming of age: Steroid hormones of wild immature baboons (*Papio cynocephalus*)." *Am J Primatol* 67(1): 83–100.

Gesquiere, L. R., E. O. Wango, et al. (2007). "Mechanisms of sexual selection: Sexual swellings and estrogen concentrations as fertility indicators and cues for male consort decisions in wild baboons." *Horm Behav* 51(1): 114–25.

Gianaros, P. J., J. A. Horenstein, et al. (2007). "Perigenual anterior cingulate morphology covaries with perceived social standing." *Soc Cogn Affect Neurosci* 2(3): 161–73.

Giedd, J. N. (2004). "Structural magnetic resonance imaging of the adolescent brain." *Ann N Y Acad Sci* 1021: 77–85.

Giedd, J. N., J. Blumenthal, et al. (1999). "Brain development during childhood and adolescence: A longitudinal MRI study." *Nat Neurosci* 2(10): 861–63.

Giedd, J. N., F. X. Castellanos, et al. (1997). "Sexual dimorphism of the developing human brain." *Prog Neuropsychopharmacol Biol Psychiatry* 21(8): 1185–1201.

Giedd, J. N., L. S. Clasen, et al. (2006). "Puberty-related influences on brain development." *Mol Cell Endocrinol* 254–55: 154–62.

Giedd, J. N., F. M. Lalonde, et al. (2009). "Anatomical brain magnetic resonance imaging of typically developing children and adolescents." *J Am Acad Child Adolesc Psychiatry* 48(5): 465–70.

Giedd, J. N., J. W. Snell, et al. (1996). "Quantitative magnetic resonance imaging of human brain development: Ages 4–18." *Cereb Cortex* 6(4): 551–60.

Giedd, J. N., A. C. Vaituzis, et al. (1996). "Quantitative MRI of the temporal lobe, amygdala, and hippocampus in normal human development: Ages 4–18 years." *J Comp Neurol* 366(2): 223–30.

Giles, J. (2006). "No such thing as excessive levels of sexual behavior." *Arch Sex Behav* 35(6): 641–42; author reply 643–44.

Gillath, O., M. Mikulincer, et al. (2008a). "When sex primes love: Subliminal sexual priming motivates relationship goal pursuit." *Pers Soc Psychol Bull* 34(8): 1057–69.

Gillath, O., P. R. Shaver, et al. (2008b). "Genetic correlates of adult attachment style." *Pers Soc Psychol Bull* 34(10): 1396–1405.

Gilmore, J. H., W. Lin, et al. (2007). "Regional gray matter growth, sexual dimorphism, and cerebral asymmetry in the neonatal brain." *J Neurosci* 27(6): 1255–60.

Ginn, S. R., and S. J. Pickens (2005). "Relationships between spatial activities and scores on the mental rotation test as a function of sex." *Perceptual and Motor Skills* 100(3): 877–81.

Ginther, A. J., A. A. Carlson, et al. (2002). "Neonatal and pubertal development in males of a cooperatively breeding primate, the cotton-top tamarin (*Saguinus oedipus oedipus*)." *Biol Reprod* 66(2): 282–90.

Gleason, E. D., M. J. Fuxjager, et al. (2009). "Testosterone release and social context: When it occurs and why." *Front Neuroendocrinol* 30(4): 460–69.

Gluckman, P. D., and M. A. Hanson (2006). "Evolution, development and timing of puberty." *Trends Endocrinol Metab* 17(1): 7–12.

Gobrogge, K. L., Y. Liu, et al. (2007). "Anterior hypothalamic neural activation and neurochemical associations with aggression in pair-bonded male prairie voles." *J Comp Neurol* 502(6): 1109–22.

Goetz, A. T. and T. K. Shackelford, eds. (2006). *Mate Retention, Semen Displacement, and Sperm Competition in Humans.* New York: Cambridge University Press.

Goetz, A. T., T. K. Shackelford, et al. (2005). "Mate retention, semen displacement, and human sperm competition: A preliminary investigation of tactics to prevent and correct female infidelity." *Personality and Individual Differences* 38(4): 749–63.

Goldin-Meadow, S., S. W. Cook, et al. (2009). "Gesturing gives children new ideas about math." *Psychol Sci* 20(3): 267–72.

Gomes, C. M., and C. Boesch (2009). "Wild chimpanzees exchange meat for sex on a long-term basis." *PLoS One* 4(4): e5116.

Gonzaga, G. C., R. A. Turner, et al. (2006). "Romantic love and sexual desire in close relationships." *Emotion* 6(2): 163–79.

Good, C. D., K. Lawrence, et al. (2003). "Dosage-sensitive X-linked locus influences the development of amygdala and orbitofrontal cortex, and fear recognition in humans." *Brain* 126(11): 2431–46.

Gooren, L. J. (2009). "Late-onset hypogonadism." *Front Horm Res* 37: 62–73.

Gooren, L. J., and H. M. Behre (2008). "Testosterone treatment of hypogonadal men participating in competitive sports." *Andrologia* 40(3): 195–99.

Gordon, I., and R. Feldman (2008). "Synchrony in the triad: A microlevel process model of coparenting and parent-child interactions." *Fam Process* 47(4): 465–79.

Gottman, J. M., J. S. Gottman, and J. Declaire (2006). *Ten Lessons to Transform Your Marriage: America's Love Lab Experts Share Their Strategies for Strengthening Your Relationship.* New York: Crown.

Goy, R. W., F. B. Bercovitch, et al. (1988). "Behavioral masculinization is independent of genital masculinization in prenatally androgenized female rhesus macaques." *Horm Behav* 22(4): 552–71.

Grafton, S. T., L. Fadiga, et al. (1997). "Premotor cortex activation during observation and naming of familiar tools." *Neuroimage* 6(4): 231–36.

Gragasin, F. S., E. D. Michelakis, et al. (2004). "The neurovascular mechanism of clitoral erection: Nitric oxide and cGMP-stimulated activation of BKCa channels." *FASEB J* 18(12): 1382–91.

Grant, L. (1985). *Race, Gender, Status, Classroom Interactions and Children's Socialization in Elementary School.* Orlando, FL: Academic Press.

Gray, P. B., J. C. Parkin, et al. (2007). "Hormonal correlates of human paternal interactions: A hospital-based investigation in urban Jamaica." *Horm Behav* 52(4): 499–507.

Gray, P. B., C. F. Yang, et al. (2006). "Fathers have lower salivary testosterone levels than unmarried men and married non-fathers in Beijing, China." *Proc Biol Sci* 273(1584): 333–39.

Graziano, W. G., L. A. Jensen-Campbell, et al., eds. (1997). *Interpersonal Attraction from an Evolutionary Psychology Perspective: Women's Reactions to Dominant and Prosocial Men.* Hillsdale, NJ: Lawrence Erlbaum Associates.

Greeno, C. G., and E. E. Maccoby, eds. (1994). *How Different is the "Different Voice"?* New York: Garland.

Grewen, K. M., S. S. Girdler, et al. (2005). "Effects of partner support on resting oxytocin, cortisol, norepinephrine, and blood pressure before and after warm partner contact." *Psychosom Med* 67(4): 531–38.

Griskevicius, V., J. M. Tybur, et al. (2009). "Aggress to impress: Hostility as an evolved context-dependent strategy." *Journal of Personality and Social Psychology* 96(5): 980–94.

Griskevicius, V., J. M. Tybur, et al. (2007). "Blatant benevolence and conspicuous consumption: When romantic motives elicit strategic costly signals." *J Pers Soc Psychol* 93(1): 85–102.

Grosbras, M.-H., M. Jansen, et al. (2007). "Neural mechanisms of resistance to peer influence in early adolescence." *Journal of Neuroscience* 27(30): 8040–45.

Grossmann, K., K. E. Grossmann, E. Fremmer-Bombik, H. Kindler, H. Scheuerer-Englisch, and P. Zimmermann (2002). "The uniqueness of the child-father attachment relationship: Fathers' sensitive and challenging play as a pivotal variable in a 16-year longitudinal study." *Social Development* 11(3): 307–31.

Grossmann, T., T. Striano, et al. (2007). "Developmental changes in infants' processing of happy and angry facial expressions: A neurobehavioral study." *Brain Cogn* 64: 30–41.

Guastella, A. J., D. S. Carson, et al. (2009). "Does oxytocin influence the early detection of angry and happy faces?" *Psychoneuroendocrinology* 34(2): 220–25.

Guastella, A. J., P. B. Mitchell, et al. (2008a). "Oxytocin enhances the encoding of positive social memories in humans." *Biol Psychiatry* 64(3): 256–58.

Guastella, A. J., P. B. Mitchell, et al. (2008b). "Oxytocin increases gaze to the eye region of human faces." *Biol Psychiatry* 63(1): 3–5.

Guiliano, F., O. Rampin, et al. (1997). "[The peripheral pharmacology of erection]." *Prog Urol* 7(1): 24–33.

Guiller, J., and A. Durndell (2007). "Students' linguistic behaviour in online discussion groups: Does gender matter?" *Computers in Human Behavior* 23(5): 2240–55.

Gurven, M., and K. Hill (2009). "Why do men hunt? A reevaluation of 'man the hunter' and the sexual division of labor." *Curr Anthropol* 50(1): 51–62; discussion 62–74.

Guyer, A. E., C. S. Monk, et al. (2008). "A developmental examination of amygdala response to facial expressions." *J Cogn Neurosci* 20(9): 1565–82.

Guzell, J. R., and L. Vernon-Feagans (2004). "Parental perceived control over caregiving and its relationship to parent-infant interaction." *Child Dev* 75(1): 134–46.

Hadler, I. (2007). "Anger and aggression may be genetic." Paper presented at American Psychosomatic Society Annual Meeting, Budapest.

Haenfler, R. (2004). "Manhood in contradiction: The two faces of straight edge." *Men and Masculinities* 7(1): 77–79.

Hagenauer, M. H., J. I. Perryman, et al. (2009). "Adolescent changes in the homeostatic and circadian regulation of sleep." *Dev Neurosci* 31(4): 276–84.

Hagiwara, H., T. Funabashi, et al. (2007). "Effects of neonatal testosterone treatment on sex differences in formalin-induced nociceptive behavior in rats." *Neurosci Lett* 412(3): 264–67.

Hahn, N., P. Jansen, et al. (2009). "Preschoolers' mental rotation: Gender differences in hemispheric asymmetry." *J Cogn Neurosci*, published online April 14, 2009.

Halari, R., M. Hines, et al. (2005). "Sex differences and individual differences in cognitive performance and their relationship to endogenous gonadal hormones and gonadotropins." *Behav Neurosci* 119(1): 104–17.

Hall, G. B., S. F. Witelson, et al. (2004). "Sex differences in functional activation patterns revealed by increased emotion processing demands." *Neuroreport* 15(2): 219–23.

Hall, J. A. (1984). *Nonverbal Sex Differences: Communication Accuracy and Expressive Style.*" Baltimore: Johns Hopkins University Press.

Hall, J. A. (1978). "Gender effects in decoding nonverbal cues." *Psychological Bulletin* 85: 845–57.

Hall, J. A., J. D. Carter, and T. G. Horgan (2000). "Gender differences in the nonverbal communication of emotion." In: A. H. Fischer, ed., *Gender and Emotion: Social Psychological Perspectives.* Paris: Cambridge University Press, pp. 97–117.

Hall, S. A., G. R. Esche, et al. (2008). "Correlates of low testosterone and symptomatic androgen deficiency in a population-based sample." *J Clin Endocrinol Metab* 93(10): 3870–77.

Halle, C., T. Dowd, et al. (2008). "Supporting fathers in the transition to parenthood." *Contemp Nurse* 31(1): 57–70.

Halpern, C. T., B. Campbell, et al. (2002). "Associations between stress reactivity and sexual and nonsexual risk taking in young adult human males." *Horm Behav* 42(4): 387–98.

Halpern, C. T., K. Joyner, et al. (2000). "Smart teens don't have sex (or kiss much either)." *J Adolesc Health* 26(3): 213–25.

Halpern, C. T., J. R. Udry, et al. (1998). "Monthly measures of salivary testosterone predict sexual activity in adolescent males." *Arch Sex Behav* 27(5): 445–65.

Halpern, C. T., J. R. Udry, et al. (1997). "Testosterone predicts initiation of coitus in adolescent females." *Psychosom Med* 59(2): 161–71.

Halpern, C. T., J. R. Udry, et al. (1994). "Testosterone and religiosity as predictors of sexual attitudes and activity among adolescent males: A biosocial model." *J Biosoc Sci* 26(2): 217–34.

Halpern, C. T., J. R. Udry, et al. (1993a). "Relationships between aggression and pubertal increases in testosterone: A panel analysis of adolescent males." *Soc Biol* 40(1–2): 8–24.

Halpern, C. T., J. R. Udry, et al. (1993b). "Testosterone and pubertal development as predictors of sexual activity: A panel analysis of adolescent males." *Psychosom Med* 55(5): 436–47.

Hamann, S. (2005). "Sex differences in the responses of the human amygdala." *Neuroscientist* 11(4): 288–93.

Hammock, E. A., and L. J. Young (2006). "Oxytocin, vasopressin and pair bonding: Implications for autism." *Philos Trans R Soc Lond B Biol Sci* 361(1476): 2187–98.

Hammond, W. P., and J. S. Mattis (2005). "Being a man about it: Manhood meaning among African American men." *Psychology of Men & Masculinity* 6(2): 114–26.

Hampson, E. (2008). "Sex differences in visuospatial perception and cognition." In: J. B. Becker, K. Berkley, N. Geary, E. Hampson, J. P. Herman, and E. A. Young, eds., *Sex Differences in the Brain: From Genes to Behavior.* Oxford, UK: Oxford University Press.

Han, T. M., and G. J. De Vries (2003). "Organizational effects of testosterone, estradiol, and dihydrotestosterone on vasopressin mRNA expression in the bed nucleus of the stria terminalis." *J Neurobiol* 54(3): 502–10.

Handa, R. J., T. R. Pak, et al. (2008). "An alternate pathway for androgen regulation of brain function: Activation of estrogen receptor beta by the metabolite of dihydrotestosterone, 5alpha-androstane-3beta,17beta-diol." *Horm Behav* 53(5): 741–52.

Hanggi, J., A. Buchmann, et al. (2009). "Sexual dimorphism in the parietal substrate associated with visuospatial cognition independent of general intelligence." *J Cogn Neurosci* 22(1): 139–55.

Harburg, E. (2008). "Marital pair anger-coping types may act as an entity to affect mortality: Preliminary findings from a prospective study (Tecumseh, Michigan, 1971–1988)." *Journal of Family Communication* 8(1): 44–61.

Harburg, E., M. Julius, et al. (2003). "Expressive/suppressive anger-coping responses, gender, and types of mortality: A 17-year follow-up (Tecumseh, Michigan, 1971–1988)." *Psychosom Med* 65(4): 588–97.

Harenski, C. L., O. Antonenko, et al. (2008). "Gender differences in neural

mechanisms underlying moral sensitivity." *Soc Cogn Affect Neurosci* 3(4): 313–21.
Hartmann, U., M. Schedlowski, et al. (2005). "Cognitive and partner-related factors in rapid ejaculation: Differences between dysfunctional and functional men." *World J Urol* 23(2): 93–101.
Haselton, M. G., D. M. Buss, et al. (2005). "Sex, lies, and strategic interference: The psychology of deception between the sexes." *Personality and Social Psychology Bulletin* 31(1): 3–23.
Haselton, M. G., and S. W. Gangestad (2006). "Conditional expression of women's desires and men's mate guarding across the ovulatory cycle." *Hormones and Behavior* 49(4): 509–18.
Hassett, J. M., E. R. Siebert, et al. (2008). "Sex differences in rhesus monkey toy preferences parallel those of children." *Horm Behav* 54(3): 359–64.
Hausmann, M., D. Schoofs, et al. (2009). "Interactive effects of sex hormones and gender stereotypes on cognitive sex differences: A psychobiosocial approach." *Psychoneuroendocrinology* 34(3): 389–401.
Havlicek, J., and S. C. Roberts (2009a). "MHC-correlated mate choice in humans: A review." *Psychoneuroendocrinology* 34(4): 497–512.
Havlicek, J., and S. C. Roberts (2009b). "Towards a neuroscience of love: Olfaction, attention and a model of neurohypophysial hormone action." *Front Evol Neurosci* 1: 2.
Hawkes, K. (2004). "Human longevity: The grandmother effect." *Nature* 428(6979): 128–29.
Hawkley, L. C., M. W. Browne, et al. (2005). "How can I connect with thee? Let me count the ways." *Psychol Sci* 16(10): 798–804.
Hawkley, L. C., M. E. Hughes, et al. (2008). "From social structural factors to perceptions of relationship quality and loneliness: The Chicago health, aging, and social relations study." *J Gerontol B Psychol Sci Soc Sci* 63(6): S375–S384.
Hawkley, L. C., C. M. Masi, et al. (2006). "Loneliness is a unique predictor of age-related differences in systolic blood pressure." *Psychol Aging* 21(1): 152–64.
Hawkley, L. C., R. A. Thisted, et al. (2009). "Loneliness predicts reduced physical activity: Cross-sectional and longitudinal analyses." *Health Psychol* 28(3): 354–63.
Hayward, C., and K. Sanborn (2002). "Puberty and the emergence of gender differences in psychopathology." *J Adolesc Health* 30(4 Suppl): 49–58.
Heaton, J. P. (2001). "Andropause: Coming of age for an old concept?" *Curr Opin Urol* 11(6): 597–601.
Heaton, J. P., and A. Morales (2001). "Andropause—a multisystem disease." *Can J Urol* 8(2): 1213–22.
Hecht, M. A., and M. LaFrance (1998). "License or obligation to smile: The

effect of power and sex on amount and type of smiling." *Personality and Social Psychology Bulletin* 24(12): 1332–42.

Heil, M., and P. Jansen-Osmann (2008). "Sex differences in mental rotation with polygons of different complexity: Do men utilize holistic processes whereas women prefer piecemeal ones?" *Q J Exp Psychol* (Colchester) 61(5): 683–89.

Heinrichs, M., and G. Domes (2008). "Neuropeptides and social behaviour: Effects of oxytocin and vasopressin in humans." *Prog Brain Res* 170: 337–50.

Hermans, E. J., N. F. Ramsey, et al. (2008). "Exogenous testosterone enhances responsiveness to social threat in the neural circuitry of social aggression in humans." *Biol Psychiatry* 63(3): 263–70.

Hermans, E. J., P. Putman, et al. (2007). "Exogenous testosterone attenuates the integrated central stress response in healthy young women." *Psychoneuroendocrinology* 32(8–10): 1052–61.

Hermans, E. J., P. Putman, et al. (2006a). "A single administration of testosterone reduces fear-potentiated startle in humans." *Biol Psychiatry* 59(9): 872–74.

Hermans, E. J., P. Putman, et al. (2006b). "Testosterone administration reduces empathetic behavior: A facial mimicry study." *Psychoneuroendocrinology* 31(7): 859–66.

Hernandez-Tristan, R., C. Arevalo, et al. (1999). "Effect of prenatal uterine position on male and female rats' sexual behavior." *Physiol Behav* 67(3): 401–8.

Herpertz, S. C., T. Vloet, et al. (2007). "Similar autonomic responsivity in boys with conduct disorder and their fathers." *J Am Acad Child Adolesc Psychiatry* 46(4): 535–44.

Herry, C., S. Ciocchi, et al. (2008). "Switching on and off fear by distinct neuronal circuits." *Nature* 454(7204): 600–606.

Herve, P. Y., G. Leonard, et al. (2009). "Handedness, motor skills and maturation of the corticospinal tract in the adolescent brain." *Hum Brain Mapp* 30(10): 3151–62.

Hill, C. A. (2002). "Gender, relationship stage, and sexual behavior: The importance of partner emotional investment within specific situations." *J Sex Res* 39(3): 228–40.

Hill, R. A., and R.I.M. Dunbar (2003). "Social network size in humans." *Human Nature* 14(1): 53–72.

Hines, M. (2002). "Sexual differentiation of human brain and behavior." In: D. W. Pfaff, ed., *Hormones, Brain and Behavior*, vol. 4, pp. 425–62.

Hines, M., S. F. Ahmed, et al. (2003). "Psychological outcomes and gender-related development in complete androgen insensitivity syndrome." *Arch Sex Behav* 32(2): 93–101.

Hines, M., L. S. Allen, et al. (1992). "Sex differences in subregions of the medial nucleus of the amygdala and the bed nucleus of the stria terminalis of the rat." *Brain Res* 579(2): 321–26.

Hines, M., C. Brook, et al. (2004). "Androgen and psychosexual development: Core gender identity, sexual orientation and recalled childhood gender role behavior in women and men with congenital adrenal hyperplasia (CAH)." *J Sex Res* 41(1): 75–81.

Hines, M., B. A. Fane, et al. (2003). "Spatial abilities following prenatal androgen abnormality: Targeting and mental rotations performance in individuals with congenital adrenal hyperplasia." *Psychoneuroendocrinology* 28(8): 1010–26.

Hines, M., and F. R. Kaufman (1994). "Androgen and the development of human sex-typical behavior: Rough-and-tumble play and sex of preferred playmates in children with congenital adrenal hyperplasia (CAH)." *Child Dev* 65(4): 1042–53.

Hiort, O., and P. M. Holterhus (2003). "Androgen insensitivity and male infertility." *Int J Androl* 26(1): 16–20.

Hiort, O., and P. M. Holterhus (2000). "The molecular basis of male sexual differentiation." *Eur J Endocrinol* 142(2): 101–10.

Hiort, O., P. M. Holterhus, et al. (2000). "Significance of mutations in the androgen receptor gene in males with idiopathic infertility." *J Clin Endocrinol Metab* 85(8): 2810–15.

Hiort, O., P. M. Holterhus, et al. (1998). "Physiology and pathophysiology of androgen action." *Baillieres Clin Endocrinol Metab* 12(1): 115–32.

Hiort, O., G. H. Sinnecker, et al. (1996). "The clinical and molecular spectrum of androgen insensitivity syndromes." *Am J Med Genet* 63(1): 218–22.

Hittelman, J.H.D. (1979). "Sex differences in neonatal eye contact time." *Merrill-Palmer Quarterly* 25: 171–84.

Hoeft, F., C. L. Watson, et al. (2008). "Gender differences in the mesocorticolimbic system during computer game-play." *J Psychiatr Res* 42(4): 253–58.

Hoehl, S., and T. Striano (2008). "Neural processing of eye gaze and threat-related emotional facial expressions in infancy." *Child Dev* 79(6): 1752–60.

Hofer, A., C. M. Siedentopf, et al. (2006). "Gender differences in regional cerebral activity during the perception of emotion: A functional MRI study." *Neuroimage* 32(2): 854–62.

Hogervorst, E., V. W. Henderson, R. B. Gibbs, and R. Diaz Brinton, eds. (2009). *Hormones, Cognition and Dementia: State of the Art and Emergent Therapeutic Strategies*. London: Cambridge University Press.

Holden, C. (2004). "An everlasting gender gap?" *Science* 305(5684): 639–40.

Holst, S., I. Lund, et al. (2005). "Massage-like stroking influences plasma levels of gastrointestinal hormones, including insulin, and increases weight gain in male rats." *Auton Neurosci* 120(1–2): 73–79.

Holstege, G., J. R. Georgiadis, et al. (2003). "Brain activation during human male ejaculation." *J Neurosci* 23(27): 9185–93.

Holt-Lunstad, J., W. A. Birmingham, et al. (2008). "Influence of a 'warm touch' support enhancement intervention among married couples on ambulatory blood pressure, oxytocin, alpha amylase, and cortisol." *Psychosom Med* 70(9): 976–85.

Honekopp, J., M. Voracek, et al. (2006). "2nd to 4th digit ratio (2D:4D) and number of sex partners: Evidence for effects of prenatal testosterone in men." *Psychoneuroendocrinology* 31(1): 30–37.

Hooven, C. K., C. F. Chabris, et al. (2004). "The relationship of male testosterone to components of mental rotation." *Neuropsychologia* 42(6): 782–90.

Hrdy, S. B. (2000). "The optimal number of fathers: Evolution, demography, and history in the shaping of female mate preferences." *Ann N Y Acad Sci* 907: 75–96.

Hrdy, S. B. (1974). "Male-male competition and infanticide among the langurs (*Presbytis entellus*) of Abu, Rajasthan." *Folia Primatol (Basel)* 22(1): 19–58.

Hu, S. H., N. Wei, et al. (2008). "Patterns of brain activation during visually evoked sexual arousal differ between homosexual and heterosexual men." *AJNR Am J Neuroradiol* 29(10): 1890–96.

Huber, D., P. Veinante, et al. (2005). "Vasopressin and oxytocin excite distinct neuronal populations in the central amygdala." *Science* 308(5719): 245–48.

Hugdahl, K., T. Thomsen, et al. (2006). "Sex differences in visuo-spatial processing: An fMRI study of mental rotation." *Neuropsychologia* 44(9): 1575–83.

Hughes, S. M., F. Dispenza, et al. (2004). "Ratings of voice attractiveness predict sexual behavior and body configuration." *Evolution and Human Behavior* 25(5): 295–304.

Hughes, S. M, G. G. Gallup, Jr. (2008). "Why are we attracted to certain voices? Voice as an evolved medium for the transmission of psychological and biological information." In: K. Izdebski, ed., *Emotions in the Human Voice*, vol. 2: Clinical Evidence. San Diego: Plural Publishing.

Hughes, S. M., M. A. Harrison, et al. (2007). "Sex differences in romantic kissing among college students: An evolutionary perspective." *Evolutionary Psychology* 5(3): 612–31.

Huh, J., K. Park, et al. (2008). "Brain activation areas of sexual arousal with olfactory stimulation in men: A preliminary study using functional MRI." *J Sex Med* 5(3): 619–25.

Hummel, T., F. Krone, et al. (2005). "Androstadienone odor thresholds in adolescents." *Horm Behav* 47(3): 306–10.

Hummer, T. A., and M. K. McClintock (2010). "Putative human pheromone androstadienone attunes the mind specifically to emotional information." *Horm Behav* 55(4): 548–59.

Hunter, A. G., C. A. Friend, et al. (2006). "Loss, survival, and redemption: African American male youths' reflections on life without fathers, manhood, and coming of age." *Youth & Society* 37(4): 423–52.

Huttenlocher, J., W. Haight, et al. (1991). "Early vocabulary growth: Relation to language input and gender." *Developmental Psychology* 27(2): 236–48.

Hyde, J. S. (2005). "The gender similarities hypothesis." *American Psychologist* 60(6): 581–92.

Iervolino, A. C., M. Hines, et al. (2005). "Genetic and environmental influences on sex-typed behavior during the preschool years." *Child Dev* 76(4): 826–40.

Iijima, M., O. Arisaka, et al. (2001). "Sex differences in children's free drawings: A study on girls with congenital adrenal hyperplasia." *Horm Behav* 40(2): 99–104.

Ikezawa, S., K. Nakagome, et al. (2008). "Gender differences in lateralization of mismatch negativity in dichotic listening tasks." *Int J Psychophysiol* 68(1): 41–50.

Immordino-Yang, M. H., A. McColl, et al. (2009). "Neural correlates of admiration and compassion." *Proc Natl Acad Sci U S A* 106(19): 8021–26.

Impett, E. A., A. Strachman, et al. (2008). "Maintaining sexual desire in intimate relationships: The importance of approach goals." *J Pers Soc Psychol* 94(5): 808–23.

Insel, T. R. (2003). "Is social attachment an addictive disorder?" *Physiol Behav* 79(3): 351–57.

Insel, T. R., and R. D. Fernald (2004). "How the brain processes social information: Searching for the social brain." *Annu Rev Neurosci* 27: 697–722.

Isidori, A. M., E. Giannetta, et al. (2005a). "Effects of testosterone on body composition, bone metabolism and serum lipid profile in middle-aged men: A meta-analysis." *Clin Endocrinol* (Oxford) 63(3): 280–93.

Isidori, A. M., E. Giannetta, et al. (2005b). "Effects of testosterone on sexual function in men: Results of a meta-analysis." *Clin Endocrinol* (Oxford) 63(4): 381–94.

Jabbi, M., and C. Keysers (2008). "Inferior frontal gyrus activity triggers anterior insula response to emotional facial expressions." *Emotion* 8(6): 775–80.

Jacklin, C., and E. Maccoby (1978). "Social behavior at thirty-three months in same-sex and mixed-sex dyads." *Child Development* 49: 557–69.

Jacklin, C. N., K. T. Wilcox, et al. (1988). "Neonatal sex-steroid hormones and cognitive abilities at six years." *Developmental Psychobiology* 21(6): 567–74.

Jacob, S., S. Garcia, et al. (2002). "Psychological effects of musky compounds: Comparison of androstadienone with androstenol and muscone." *Horm Behav* 42(3): 274–83.

Jamin, S. P., N. A. Arango, et al. (2002). "Genetic studies of MIS signalling in sexual development." *Novartis Found Symp* 244: 157–64; discussion 164–68, 203–6, 253–57.

Janssen, E., K. R. McBride, et al. (2008). "Factors that influence sexual arousal in men: A focus group study." *Arch Sex Behav* 37(2): 252–65.

Jarvinen, D. W., and J. G. Nicholls (1996). "Adolescents' social goals, beliefs about the causes of social success, and satisfaction in peer relations." *Developmental Psychology* 32(3): 435–41.

Jensen-Campbell, L. A., W. G. Graziano, et al. (1995). "Dominance, prosocial orientation, and female preferences: Do nice guys really finish last?" *Journal of Personality and Social Psychology* 68(3): 427–40.

Jiang, J., U. Rosenqvist, et al. (2007). "Influence of grandparents on eating behaviors of young children in Chinese three-generation families." *Appetite* 48(3): 377–83.

Jocham, G., J. Neumann, et al. (2009). "Adaptive coding of action values in the human rostral cingulate zone." *J Neurosci* 29(23): 7489–96.

Johnson, D. D., R. McDermott, et al. (2006). "Overconfidence in wargames: Experimental evidence on expectations, aggression, gender and testosterone." *Proc Biol Sci* 273(1600): 2513–20.

Johnson, E. O., T. Roth, et al. (2006). "Epidemiology of DSM-IV insomnia in adolescence: Lifetime prevalence, chronicity, and an emergent gender difference." *Pediatrics* 117(2): e247–56.

Jones, B. C., L. M. DeBruine, et al. (2008). "Effects of menstrual cycle phase on face preferences." *Arch Sex Behav* 37(1): 78–84.

Jones, B. C., D. R. Feinberg, et al. (2008). "Integrating cues of social interest and voice pitch in men's preferences for women's voices." *Biol Lett* 4(2): 192–94.

Jordan, C. L., and L. Doncarlos (2008). "Androgens in health and disease: An overview." *Horm Behav* 53(5): 589–95.

Jordan, K., T. Wustenberg, et al. (2002). "Women and men exhibit different cortical activation patterns during mental rotation tasks." *Neuropsychologia* 40(13): 2397–2408.

Josephs, R. A., H. R. Markus, et al. (1992). "Gender and self-esteem." *J Pers Soc Psychol* 63(3): 391–402.

Josephs, R. A., M. L. Newman, et al. (2003). "Status, testosterone, and human intellectual performance: Stereotype threat as status concern." *Psychol Sci* 14(2): 158–63.

Josephs, R. A., J. G. Sellers, et al. (2006). "The mismatch effect: When testosterone and status are at odds." *J Pers Soc Psychol* 90(6): 999–1013.

Josso, N., I. Lamarre, et al. (1993). "Anti-Müllerian hormone in early human development." *Early Hum Dev* 33(2): 91–99.

Juntti, S. A., J. K. Coats, et al. (2008). "A genetic approach to dissect sexually dimorphic behaviors." *Horm Behav* 53(5): 627–37.

Kahlenberg, S. M., M. E. Thompson, et al. (2008). "Immigration costs for female chimpanzees and male protection as an immigrant counterstrategy to intrasexual aggression." *Animal Behaviour* 76(5): 1497–1509.

Kahnt, T., S. Q. Park, et al. (2009). "Dorsal striatal-midbrain connectivity in humans predicts how reinforcements are used to guide decisions." *J Cogn Neurosci* 21(7): 1332–45.

Kaighobadi, F., and T. K. Shackelford (2008). "Female attractiveness mediates the relationship between in-pair copulation frequency and men's mate retention behaviors." *Personality and Individual Differences* 45(4): 293–95.

Kaighobadi, F., V. G. Starratt, et al. (2008). "Male mate retention mediates the relationship between female sexual infidelity and female-directed violence." *Personality and Individual Differences* 44(6): 1422–31.

Kaiser, S., S. Walther, et al. (2008). "Gender-specific strategy use and neural correlates in a spatial perspective taking task." *Neuropsychologia* 46(10): 2524–31.

Kajantie, E. D. Phillips. (2006). "The effects of sex and hormonal status on the physiological response to acute psychosocial stress." *Psychoneuroendocrinology* 31 (2): 151–78.

Kaplan, H. (1997). "The evolution of the human life course." In: K. W. Wachter and C. E. Finch, eds., *Between Zeus and the Salmon.* Washington, DC: National Academy Press, 175–211.

Kauffman, A. S., V. M. Navarro, et al. (2010). "Sex differences in the regulation of kiss1/NKB neurons in juvenile mice: Implications for the timing of puberty." *Am J Physiol Endocrinol Metab* 297: E1212–21.

Kaufman, J. M., and A. Vermeulen (2005). "The decline of androgen levels in elderly men and its clinical and therapeutic implications." *Endocr Rev* 26(6): 833–76.

Keating, D. P., ed. (2004). *Cognitive and brain development.* Hoboken, NJ: John Wiley.

Keller, K., and V. Menon (2009). "Gender differences in the functional and structural neuroanatomy of mathematical cognition." *Neuroimage* 47(1): 342–52.

Kendall, S., and D. Tannen, eds. (1997). *Gender and Language in the Workplace.* Thousand Oaks, CA: Sage Publications.

Kendrick, K. M. (2000). "Oxytocin, motherhood and bonding." *Exp Physiol* 85: 111S-124S.

Keverne, E. B. (2007). "Genomic imprinting and the evolution of sex differences in mammalian reproductive strategies." *Adv Genet* 59: 217–43.

Keverne, E. B. (2004a). "Importance of olfactory and vomeronasal systems for male sexual function." *Physiol Behav* 83(2): 177–87.

Keverne, E. B., and J. P. Curley (2004b). "Vasopressin, oxytocin and social behaviour." *Curr Opin Neurobiol* 14(6): 777–83.

Kiecolt-Glaser, J. K., R. Glaser, et al. (1998). "Marital stress: Immunologic, neuroendocrine, and autonomic correlates." *Ann N Y Acad Sci* 840: 656–63.

Kiecolt-Glaser, J. K., T. J. Loving, et al. (2005). "Hostile marital interactions, proinflammatory cytokine production, and wound healing." *Arch Gen Psychiatry* 62(12): 1377–84.

Kiecolt-Glaser, J. K., and T. L. Newton (2001). "Marriage and health: His and hers." *Psychol Bull* 127(4): 472–503.

Kiecolt-Glaser, J. K., T. Newton, et al. (1996). "Marital conflict and endocrine function: Are men really more physiologically affected than women?" *J Consult Clin Psychol* 64(2): 324–32.

Kilpatrick, L. A., D. H. Zald, et al. (2006). "Sex-related differences in amygdala functional connectivity during resting conditions." *Neuroimage* 30(2): 452–61.

Kimchi, T., J. Xu, et al. (2007). "A functional circuit underlying male sexual behaviour in the female mouse brain." *Nature* 448(7157): 1009–14.

Kimura, K., T. Hachiya, et al. (2008). "Fruitless and doublesex coordinate to generate male-specific neurons that can initiate courtship." *Neuron* 59(5): 759–69.

King, S. R. (2008). "Emerging roles for neurosteroids in sexual behavior and function." *J Androl* 29(5): 524–33.

King, V., and G. H. Elder, Jr. (1998a). "Education and grandparenting roles." *Research on Aging* 20(4): 450–74.

King, V., and G. H. Elder Jr. (1997). "The legacy of grandparenting: Childhood experiences with grandparents and current involvement with grandchildren." *Journal of Marriage & the Family* 59(4): 848–59.

King, V., and G. H. Elder Jr. (1995). "American children view their grandparents: Linked lives across three rural generations." *Journal of Marriage & the Family* 57(1): 165–78.

King, V., S. T. Russell, et al., eds. (1998b). *Grandparenting in family systems: An ecological perspective.* Westport, CT: Greenwood.

King, V., M. Silverstein, et al. (2003). "Relations with grandparents: Rural Midwest versus urban southern California." *Journal of Family Issues* 24(8): 1044–69.

Kinnunen, L. H., H. Moltz, et al. (2004). "Differential brain activation in exclusively homosexual and heterosexual men produced by the selective serotonin reuptake inhibitor, fluoxetine." *Brain Res* 1024(1–2): 251–54.

Kinsley, C. H., and K. G. Lambert (2008). "Reproduction-induced neuroplasticity: Natural behavioural and neuronal alterations associated with the production and care of offspring." *J Neuroendocrinol* 20(4): 515–25.

Kirk, K. M., J. M. Bailey, et al. (2000). "Etiology of male sexual orientation in an Australian twin sample." *Psychology, Evolution & Gender* 2(3): 301–11.

Kirsch, P., C. Esslinger, et al. (2005). "Oxytocin modulates neural circuitry for social cognition and fear in humans." *J Neurosci* 25(49): 11489–93.

Kivett, V. R., ed. (1998). *Transitions in Grandparents' Lives: Effects on the Grandparent Role.* Westport, CT: Greenwood.

Klapwijk, A., and P. A. Van Lange (2009). "Promoting cooperation and trust in 'noisy' situations: The power of generosity." *J Pers Soc Psychol* 96(1): 83–103.

Klein, H. (1991). "Couvade syndrome: Male counterpart to pregnancy." *Int J Psychiatry Med* 21(1): 57–69.

Klein, K. O., P. M. Martha Jr., et al. (1996). "A longitudinal assessment of hormonal and physical alterations during normal puberty in boys, pt. 2: Estrogen levels as determined by an ultrasensitive bioassay." *J Clin Endocrinol Metab* 81(9): 3203–7.

Klucharev, V., K. Hytonen, et al. (2009). "Reinforcement learning signal predicts social conformity." *Neuron* 61(1): 140–51.

Klusmann, D. (2002). "Sexual motivation and the duration of partnership." *Arch Sex Behav* 31(3): 275–87.

Knickmeyer, R. C., and S. Baron-Cohen (2006a). "Fetal testosterone and sex differences." *Early Hum Dev* 82(12): 755–60.

Knickmeyer, R. C., and S. Baron-Cohen (2006b). "Fetal testosterone and sex differences in typical social development and in autism." *J Child Neurol* 21(10): 825–45.

Knickmeyer, R., S. Baron-Cohen, et al. (2006c). "Fetal testosterone and empathy." *Horm Behav* 49(3): 282–92.

Knickmeyer, R., S. Baron-Cohen, et al. (2005). "Foetal testosterone, social relationships, and restricted interests in children." *J Child Psychol Psychiatry* 46(2): 198–210.

Knickmeyer, R. C., S. Wheelwright, et al. (2005). "Gender-typed play and amniotic testosterone." *Dev Psychol* 41(3): 517–28.

Knutson, B., and S. M. Greer (2008). "Anticipatory affect: Neural correlates and consequences for choice." *Philos Trans R Soc Lond B Biol Sci* 363(1511): 3771–86.

Kontula, O., and E. Haavio-Mannila (2009). "The impact of aging on human sexual activity and sexual desire." *Journal of Sex Research* 46(1): 46–56.

Kontula, O., and E. Haavio-Mannila (2002). "Masturbation in a generational perspective." *Journal of Psychology & Human Sexuality* 14(2–3): 49–83.

Kontula, O., and E. Haavio-Mannila (1994). "Sexual behavior changes in Finland during the last 20 years." *Nordisk Sexologi* 12(3): 196–214.

Korkmaz Cetin, S., T. Bildik, et al. (2008). "[Sexual behavior and sources of information about sex among male adolescents: An 8-year follow-up]." *Turk Psikiyatri Derg* 19(4): 390–97.

Korzan, W. J., G. L. Forster, et al. (2006). "Dopaminergic activity modulation via aggression, status, and a visual social signal." *Behav Neurosci* 120(1): 93–102.

Koscik, T., D. O'Leary, et al. (2009). "Sex differences in parietal lobe morphology: Relationship to mental rotation performance." *Brain Cogn* 69(3): 451–59.

Kosfeld, M., M. Heinrichs, et al. (2005). "Oxytocin increases trust in humans." *Nature* 435(7042): 673–76.

Kozorovitskiy, Y., and E. Gould (2004). "Dominance hierarchy influences adult neurogenesis in the dentate gyrus." *J Neurosci* 24(30): 6755–59.

Kozorovitskiy, Y., C. G. Gross, et al. (2005). "Experience induces structural and biochemical changes in the adult primate brain." *Proc Natl Acad Sci U S A* 102(48): 17478–82.

Kozorovitskiy, Y., M. Hughes, et al. (2006). "Fatherhood affects dendritic spines and vasopressin V1a receptors in the primate prefrontal cortex." *Nat Neurosci* 9(9): 1094–95.

Kraemer, W. J., D. N. French, et al. (2004). "Changes in exercise performance and hormonal concentrations over a Big Ten soccer season in starters and nonstarters." *J Strength Cond Res* 18(10): 121–28.

Kranz, F., and A. Ishai (2006). "Face perception is modulated by sexual preference." *Curr Biol* 16(1): 63–68.

Kringelbach, M. L., A. Lehtonen, et al. (2008). "A specific and rapid neural signature for parental instinct." *PLoS One* 3(2): e1664.

Krist, H. (2003). "Knowing how to project objects." *Journal of Cognition and Development* 4(4): 383–414.

Kruger, T., M. S. Exton, et al. (1998). "Neuroendocrine and cardiovascular response to sexual arousal and orgasm in men." *Psychoneuroendocrinology* 23(4): 401–11.

Kruger, T. H., U. Hartmann, et al. (2005). "Prolactinergic and dopaminergic mechanisms underlying sexual arousal and orgasm in humans." *World J Urol* 23(2): 130–38.

Kuzawa, C. W., L. Gettler, et al. (2009). "Fatherhood, pairbonding and testosterone in the Philippines." *Horm Behav* 56(4): 429–35.

Lamb, M. (1981). *The Role of the Father in Child Development*. New York: Wiley.

Lamm, C., C. D. Batson, et al. (2007). "The neural substrate of human empathy: Effects of perspective-taking and cognitive appraisal." *J Cogn Neurosci* 19(1): 42–58.

Lamm, C., M. H. Fischer, et al. (2007). "Predicting the actions of others taps into one's own somatosensory representations: A functional MRI study." *Neuropsychologia* 45(11): 2480–91.

Lamm, C., A. N. Meltzoff, et al. (2010). "How do we empathize with someone who is not like us? A functional magnetic resonance imaging study." *J Cogn Neurosci* 0:0, 362–76.

Lampert, M. D., and S. M. Ervin-Tripp (2006). "Risky laughter: Teasing and self-directed joking among male and female friends." *Journal of Pragmatics* 38(1): 51–72.

Langstrom, N., and R. K. Hanson (2006). "High rates of sexual behavior in the general population: Correlates and predictors." *Arch Sex Behav* 35(1): 37–52.

Lapauw, B., S. Goemaere, et al. (2008). "The decline of serum testosterone levels in community-dwelling men over 70 years of age: Descriptive data and predictors of longitudinal changes." *Eur J Endocrinol* 159(4): 459–68.

Larsen, C. M., I. C. Kokay, et al. (2008). "Male pheromones initiate prolactin-induced neurogenesis and advance maternal behavior in female mice." *Horm Behav* 53(4): 509–17.

Larsen, P. R., ed. (2003). *Williams Textbook of Endocrinology*, 10th ed. 2003.

Laughlin, G. A., E. Barrett-Connor, et al. (2008). "Low serum testosterone and mortality in older men." *J Clin Endocrinol Metab* 93(1): 68–75.

Laumann, E. O., A. Paik, et al. (1999a). "The epidemiology of erectile dysfunction: Results from the National Health and Social Life Survey." *Int J Impot Res* 11, suppl. 1: S60–S64.

Laumann, E. O., A. Paik, et al. (1999b). "Sexual dysfunction in the United States: Prevalence and predictors." *JAMA* 281(6): 537–44.

Lavelli, M., and A. Fogel (2002). "Developmental changes in mother-infant face-to-face communication: Birth to 3 months." *Dev Psychol* 38(2): 288–305.

Lavranos, G., R. Angelopoulou, et al. (2006). "Hormonal and meta-hormonal determinants of sexual dimorphism." *Coll Antropol* 30(3): 659–63.

Leal, N. L., and N. A. Pachana (2008). "Adapting the propensity for angry driving scale for use in Australian research." *Accid Anal Prev* 40(6): 2008–14.

Leaper, C., and M. M. Ayres (2007). "A meta-analytic review of gender variations in adults' language use: Talkativeness, affiliative speech, and assertive speech." *Pers Soc Psychol Rev* 11(4): 328–63.

Leaper, C. E. (2002). "Parenting girls and boys." In: Handbook of Parenting, vol. 1, *Children and Parenting*, 2 ed. Mahwah, NJ: Lawrence Erlbaum Associates.

Leckman, J. F., R. Feldman, et al. (2004). "Primary parental preoccupation: Circuits, genes, and the crucial role of the environment." *J Neural Transm* 111(7): 753–71.

Lee, A. W., N. Devidze, et al. (2006). "Functional genomics of sex hormone–dependent neuroendocrine systems: Specific and generalized actions in the CNS." *Prog Brain Res* 158: 243–72.

Lee, M. M., P. K. Donahoe, et al. (1996). "Müllerian inhibiting substance in humans: Normal levels from infancy to adulthood." *J Clin Endocrinol Metab* 81(2): 571–76.

Lee, P. A., R. K. Danish, et al. (1980). "Micropenis, pt. 3: Primary hypogonadism, partial androgen insensitivity syndrome, and idiopathic disorders." *Johns Hopkins Med J* 147(5): 175–81.

Lee, P. A., T. Mazur, et al. (1980). "Micropenis, pt. 1: Criteria, etiologies and classification." *Johns Hopkins Med J* 146(4): 156–63.

Leeb, R.T.R., and F. Gillian (2004). "Here's looking at you, kid! A longitudinal study of perceived gender differences in mutual gaze behavior in young infants." *Sex Roles* 50(1–2): 1–5

Lehman, P. (1993). *Running Scared: Masculinity and the Representation of the Male Body*. Philadelphia: Temple University Press.

Lenroot, R. K., and J. N. Giedd (2008). "The changing impact of genes and environment on brain development during childhood and adolescence: Initial findings from a neuroimaging study of pediatric twins." *Dev Psychopathol* 20(4): 1161–75.

Lenroot, R. K., and J. N. Giedd (2006). "Brain development in children and adolescents: Insights from anatomical magnetic resonance imaging." *Neurosci Biobehav Rev* 30(6): 718–29.

Lenroot, R. K., N. Gogtay, et al. (2007). "Sexual dimorphism of brain developmental trajectories during childhood and adolescence." *Neuroimage* 36(4): 1065–73.

Lenroot, R. K., J. E. Schmitt, et al. (2009). "Differences in genetic and environmental influences on the human cerebral cortex associated with development during childhood and adolescence." *Hum Brain Mapp* 30(1): 163–74.

Leppänen, J.M.H., and K. Jari (2001). "Emotion recognition and social adjustment in school-aged girls and boys." *Scandinavian Journal of Psychology* 42(5): 429–35.

Leranth, C. (2008). "Sex differences in neuroplasticity." In: J. B. Becker, K. Berkley, N. Geary, E. Hampson, J. P. Herman, and E. A. Young, eds., *Sex Differences in the Brain: From Genes to Behavior*, Oxford, UK: Oxford University Press.

LeVay, S. (1991). "A difference in hypothalamic structure between heterosexual and homosexual men." *Science* 253(5023): 1034–37.

LeVay, S., and J. Baldwin (2009). *Human sexuality*, 3 ed. Sunderland, MA: Sinauer Associates.

Levenson, R. W., L. L. Carstensen, et al. (1993). "Long-term marriage: Age, gender, and satisfaction." *Psychology and Aging* 8(2): 301–13.

Lever, J. (1976). "Sex differences in the games children play." *Social Problems* 23: 478–87.

Lever, J., D. A. Frederick, et al. (2006). "Does size matter? Men's and women's views on penis size across the lifespan." *Psychology of Men & Masculinity* 7(3): 129–43.

Levine, S. C., J. Huttenlocher, et al. (1999). "Early sex differences in spatial skill." *Developmental Psychology* 35(4): 940–49.

Levinson, D. J. (1978). *Seasons of a Man's Life*. New York: Ballantine.

Li, A. A., M. J. Baum, et al. (2008). "Building a scientific framework for studying hormonal effects on behavior and on the development of the sexually dimorphic nervous system." *Neurotoxicology* 29(3): 504–19.

Li, H., S. Pin, et al. (2005). "Sex differences in cell death." *Ann Neurol* 58(2): 317–21.

Li, W., I. Moallem, et al. (2007). "Subliminal smells can guide social preferences." *Psychol Sci* 18(12): 1044–49.

Lim, M. M., E. A. Hammock, et al. (2004a). "The role of vasopressin in the genetic and neural regulation of monogamy." *J Neuroendocrinol* 16(4): 325–32.

Lim, M. M., A. Z. Murphy, et al. (2004b). "Ventral striatopallidal oxytocin and vasopressin V1a receptors in the monogamous prairie vole (*Microtus ochrogaster*)." *J Comp Neurol* 468(4): 555–70.

Lim, M. M., Z. Wang, et al. (2004c). "Enhanced partner preference in a promiscuous species by manipulating the expression of a single gene." *Nature* 429(6993): 754–57.

Lim, M. M., and L. J. Young (2006). "Neuropeptidergic regulation of affiliative behavior and social bonding in animals." *Horm Behav* 50(4): 506–17.

Lim, M. M., and L. J. Young (2004d). "Vasopressin-dependent neural circuits underlying pair bond formation in the monogamous prairie vole." *Neuroscience* 125(1): 35–45.

Lincoln, G. A. (2001). "The irritable male syndrome." *Reprod Fertil Dev* 13(7–8): 567–76.

Lindenfors, P. (2005). "Neocortex evolution in primates: The 'social brain' is for females." *Biol Lett* 1(4): 407–10.

Lindenfors, P., L. Froberg, et al. (2004). "Females drive primate social evolution." *Proc Biol Sci* 271, suppl. 3: S101–S103.

Lindenfors, P., C. L. Nunn, et al. (2007). "Primate brain architecture and selection in relation to sex." *BMC Biol* 5: 20.

Little, A. C., B. C. Jones, et al. (2007). "Preferences for masculinity in male bodies change across the menstrual cycle." *Hormones and Behavior* 51(5): 633–39.

Liu, Y., J. T. Curtis, et al. (2001). "Vasopressin in the lateral septum regulates pair bond formation in male prairie voles (*Microtus ochrogaster*)." *Behav Neurosci* 115(4): 910–19.

Liu, Y., and Z. X. Wang (2003). "Nucleus accumbens oxytocin and dopamine interact to regulate pair bond formation in female prairie voles." *Neuroscience* 121(3): 537–44.

Lonstein, J. S., B. D. Rood, et al. (2005). "Unexpected effects of perinatal gonadal hormone manipulations on sexual differentiation of the extrahypothalamic arginine-vasopressin system in prairie voles." *Endocrinology* 146(3): 1559–67.

Lorey, B., M. Bischoff, et al. (2009). "The embodied nature of motor imagery: The influence of posture and perspective." *Exp Brain Res* 194(2): 233–43.

Lourenco, S. F., and J. Huttenlocher (2008). "The representation of geometric cues in infancy." *Infancy* 13(2): 103–27.

Loving, T. J., M.E.J. Gleason, et al. (2009). "Transition novelty moderates daters' cortisol responses when talking about marriage." *Personal Relationships* 16(2): 187–203.

Lu, S., N. G. Simon, et al. (1999). "Neural androgen receptor regulation: Effects of androgen and antiandrogen." *J Neurobiol* 41(4): 505–12.

Lu, S. F., Q. Mo, et al. (2003). "Dehydroepiandrosterone upregulates neural androgen receptor level and transcriptional activity." *J Neurobiol* 57(2): 163–71.

Luders, E. (2006). "Gender effects on cortical thickness and the influence of scaling." *Human Brain Mapping* 27(4): 314–24.

Luna, B. (2004a). "Algebra and the adolescent brain." *Trends Cogn Sci* 8(10): 437–39.

Luna, B., K. E. Garver, et al. (2004b). "Maturation of cognitive processes from late childhood to adulthood." *Child Dev* 75(5): 1357–72.

Lundberg, U. (1983). "Sex differences in behaviour pattern and catecholamine and cortisol excretion in 3–6 year old day-care children." *Biol Psychol* 16(1–2): 109–17.

Lundstrom, J. N., M. K. McClintock, et al. (2006). "Effects of reproductive state on olfactory sensitivity suggest odor specificity." *Biol Psychol* 71(3): 244–47.

Lutchmaya, S., and S. Baron-Cohen (2002a). "Human sex differences in social and non-social looking preferences, at 12 months of age." *Infant Behavior & Development* 25(3): 319–25.

Lutchmaya, S., S. Baron-Cohen, et al. (2002b). "Foetal testosterone and eye contact in 12-month-old human infants." *Infant Behavior & Development* 25(3): 327–35.

Lutchmaya, S., S. Baron-Cohen, et al. (2002c). "Foetal testosterone and vocabulary size in 18- and 24-month-old infants." *Infant Behavior & Development* 24(4): 418–24.

Lykins, A. D., M. Meana, et al. (2008). "Sex differences in visual attention to erotic and non-erotic stimuli." *Arch Sex Behav* 37(2): 219–28.

Lytton, H. R., and M. David (1991). "Parents' differential socialization of boys and girls: A meta-analysis." *Psychological Bulletin* 109(2): 267–96.

Ma, E., J. Lau, et al. (2005). "Male and female prolactin receptor mRNA expression in the brain of a biparental and a uniparental hamster, *Phodopus*, before and after the birth of a litter." *J Neuroendocrinol* 17(2): 81–90.

Maccoby, E. E. (1990). *The Role of Gender Identity and Gender Constancy in Sex-Differentiated Development.* San Francisco: Jossey-Bass.

Maccoby, E. E., ed. (2004). *Aggression in the Context of Gender Development.* New York: Guilford Publications.

Maccoby, E. E., ed. (2003). *The Gender of Child and Parent as Factors in Family Dynamics.* Mahwah, NJ: Lawrence Erlbaum Associates.

Maccoby, E. E., ed. (2002a). *The Intersection of Nature and Socialization in Childhood Gender Development.* Florence, KY: Taylor & Frances/Routledge.

Maccoby, E. E., ed. (2002b). *Perspectives on Gender Development.* New York: Psychology Press.

Maccoby, E. E. (1998). *The Two Sexes: Growing Up Apart, Coming Together.* Cambridge, MA: Harvard University Press.

Maccoby, E. E., ed. (1995). *The Two Sexes and Their Social Systems.* Washington, DC: American Psychological Association.

Maccoby, E. E. (1991). *Different Reproductive Strategies in Males and Females.* UK: Blackwell.

Maccoby, E. E., and C. N. Jacklin (1987). "Gender segregation in childhood." *Adv Child Dev Behav* 20: 239–87.

Maccoby, E. E., and C. N. Jacklin (1974). *The Psychology of Sex Differences.* Palo Alto, CA: Stanford University Press.

Maccoby, E. E., and C. N. Jacklin (1973). "Stress, activity, and proximity seeking: Sex differences in the year-old child." *Child Dev* 44(1): 34–42.

MacDonald, G., and M. R. Leary (2005). "Why does social exclusion hurt? The relationship between social and physical pain." *Psychological Bulletin* 131(2): 202–23.

Maestripieri, D., J. R. Roney, et al. (2004). "Father absence, menarche and interest in infants among adolescent girls." *Dev Sci* 7(5): 560–66.

Mak, A. K., Z. G. Hu, et al. (2010). "Sex-related differences in neural activity during emotion regulation." *Neuropsychologia* 47(13): 2900–8.

Malorni, W., I. Campesi, et al. (2007). "Redox features of the cell: A gender perspective." *Antioxid Redox Signal* 9(11): 1779–1801.

Manasco, P. K., D. M. Umbach, et al. (1995). "Ontogeny of gonadotropin, testosterone, and inhibin secretion in normal boys through puberty based on overnight serial sampling." *J Clin Endocrinol Metab* 80(7): 2046–52.

Maner, J. K., C. N. DeWall, et al. (2007a). "Does social exclusion motivate interpersonal reconnection? Resolving the 'porcupine problem.'" *J Pers Soc Psychol* 92(1): 42–55.

Maner, J. K., C. N. DeWall, et al. (2008a). "Selective attention to signs of success: Social dominance and early stage interpersonal perception." *Pers Soc Psychol Bull* 34(4): 488–501.

Maner, J. K., M. T. Gailliot, et al. (2007b). "Can't take my eyes off you: Attentional adhesion to mates and rivals." *J Pers Soc Psychol* 93(3): 389–401.

Maner, J. K., M. T. Gailliot, et al. (2007c). "Power, risk, and the status quo: Does power promote riskier or more conservative decision making?" *Pers Soc Psychol Bull* 33(4): 451–62.

Maner, J. K., S. L. Miller, et al. (2008b). "Submitting to defeat: Social anxiety, dominance threat, and decrements in testosterone." *Psychol Sci* 19(8): 764–68.

Manning, J. T. (2007). "The androgen receptor gene: A major modifier of speed of neuronal transmission and intelligence?" *Med Hypotheses* 68(4): 802–4.

Manning, J. T., B. Fink, et al. (2006). "The second to fourth digit ratio and asymmetry." *Ann Hum Biol* 33(4): 480–92.

Manning, T. (2004). "Prenatal testosterone in mind: Amniotic fluid studies." In: S. Baron-Cohen, S. Lutchmaya, and R. Knickmeyer. Cambridge, MA: MIT Press, p. 131.

Manolakou, P., G. Lavranos, et al. (2006). "Molecular patterns of sex determination in the animal kingdom: A comparative study of the biology of reproduction." *Reprod Biol Endocrinol* 4: 59.

Manoli, D. S., G. W. Meissner, et al. (2006). "Blueprints for behavior: Genetic specification of neural circuitry for innate behaviors." *Trends Neurosci* 29(8): 444–51.

Manson, J. E. (2008). "Prenatal exposure to sex steroid hormones and behavioral/cognitive outcomes." *Metabolism* 57, suppl. 2: S16–S21.

Martini, M., G. Di Sante, et al. (2008). "Androgen receptors are required for full masculinization of nitric oxide synthase system in rat limbic-hypothalamic region." *Horm Behav* 54(4): 557–64.

Masten, C. L., N. I. Eisenberger, et al. (2009). "Neural correlates of social exclusion during adolescence: Understanding the distress of peer rejection." *Soc Cogn Affect Neurosci* 4(2): 143–57.

Mather, M., and L. L. Carstensen (2005). "Aging and motivated cognition: The positivity effect in attention and memory." *Trends Cogn Sci* 9(10): 496–502.

Matsuda, K., H. Sakamoto, et al. (2008). "Androgen action in the brain and spinal cord for the regulation of male sexual behaviors." *Curr Opin Pharmacol* 8(6): 747–51.

Matthiesen, A. S., A. B. Ransjo-Arvidson, et al. (2001). "Postpartum maternal oxytocin release by newborns: Effects of infant hand massage and sucking." *Birth* 28(1): 13–19.

Mazur, A., and A. Booth (1998). "Testosterone and dominance in men." *Behav Brain Sci* 21(3): 353–63; discussion 363–97.

McCall, K. M., A. H. Rellini, et al. (2007). "Sex differences in memory for sexually relevant information." *Arch Sex Behav* 36(4): 508–17.

McCarthy, M. M. (2008). "Sex differences in the brain." In: J. B. Becker, K. Berkley, N. Geary, E. Hampson, J. P. Herman, and E. A. Young, eds., *Sex Differences in the Brain: From Genes to Behavior.* Oxford, UK: Oxford University Press.

McCarthy, M. M., G. J. De Vries, and N. G. Forger, (2009a). "Sexual differentiation of the brain: Mode, mechanisms, and meaning." In: R. H. Rubin and D. W. Pfaff, eds., *Hormones, Brain, and Behavior.* Amsterdam: Elsevier.

McCarthy, M. M., C. L. Wright, et al. (2009b). "New tricks by an old dogma: Mechanisms of the organizational/activational hypothesis of steroid-mediated sexual differentiation of brain and behavior." *Horm Behav* 55(5): 655–65.

McClure, E. B. (2000). "A meta-analytic review of sex differences in facial expression processing and their development in infants, children, and adolescents." *Psychol Bull* 126(3): 424–53.

McClure, E. B., C. S. Monk, et al. (2004). "A developmental examination of gender differences in brain engagement during evaluation of threat." *Biol Psychiatry* 55(11): 1047–55.

McCormick, C. M., and S. F. Witelson (1994). "Functional cerebral asymmetry and sexual orientation in men and women." *Behav Neurosci* 108(3): 525–31.

McCrae, R. R., and P. T. Costa Jr. (1996). *Toward a New Generation of Personality Theories: Theoretical Contexts for the Five-Factor Model.* New York: Guilford Press.

McElwain, N. L., A. G. Halberstadt, et al. (2007). "Mother- and father-reported reactions to children's negative emotions: Relations to young children's emotional understanding and friendship quality." *Child Dev* 78(5): 1407–25.

McEwen, B. S. (2009). "Introduction: The end of sex as we once knew it." *Physiol Behav* 97(2): 143–45.

McGlone, F., A. B. Vallbo, et al. (2007). "Discriminative touch and emotional touch." *Can J Exp Psychol* 61(3): 173–83.

McIntyre, M., S. W. Gangestad, et al. (2006). "Romantic involvement often reduces men's testosterone levels—but not always: The moderating role of extrapair sexual interest." *Journal of Personality and Social Psychology* 91(4): 642–51.

McKenna, K. E. (2000). "The neural control of female sexual function." *NeuroRehabilitation* 15(2): 133–43.

Meaney, M. J., and M. Szyf (2005). "Environmental programming of stress responses through DNA methylation: Life at the interface between a dynamic environment and a fixed genome." *Dialogues Clin Neurosci* 7(2): 103–23.

Mehl, M. R., S. Vazire, et al. (2007). "Are women really more talkative than men?" *Science* 317(5834): 82.

Mehta, P. H., A. C. Jones, et al. (2008). "The social endocrinology of dominance: Basal testosterone predicts cortisol changes and behavior following victory and defeat." *J Pers Soc Psychol* 94(6): 1078–93.

Mehta, P. H., E. V. Wuehrmann, et al. (2009). "When are low testosterone levels advantageous? The moderating role of individual versus intergroup competition." *Horm Behav* 56(1): 158–62.

Merzenich, M. M., J. H. Kaas, et al. (1983). "Topographic reorganization of somatosensory cortical areas 3b and 1 in adult monkeys following restricted deafferentation." *Neuroscience* 8(1): 33–55.

Meston, C. M., R. J. Levin, et al. (2004). "Women's orgasm." *Annu Rev Sex Res* 15: 173–257.

Meyer-Lindenberg, A. (2008). "Impact of prosocial neuropeptides on human brain function." *Prog Brain Res* 170: 463–70.

Miller, D. L., and L. Karakowsky (2005). "Gender influences as an impediment to knowledge sharing: When men and women fail to seek peer feedback." *J Psychol* 139(2): 101–18.

Milsted, A., L. Serova, et al. (2004). "Regulation of tyrosine hydroxylase gene transcription by Sry." *Neurosci Lett* 369(3): 203–7.

Miner, E. J., T. K. Shackelford, et al. (2009). "Mate value of romantic partners predicts men's partner-directed verbal insults." *Personality and Individual Differences* 46(2): 135–39.

Miner, E. J., V. G. Starratt, et al. (2009). "It's not all about her: Men's mate value and mate retention." *Personality and Individual Differences* 47(3): 214–18.

Minton, C., J. Kagan, et al. (1971). "Maternal control and obedience in the two-year-old." *Child Development:* 1873–94.

Miyagawa, Y., A. Tsujimura, et al. (2007). "Differential brain processing of audiovisual sexual stimuli in men: Comparative positron emission tomography study of the initiation and maintenance of penile erection during sexual arousal." *Neuroimage* 36(3): 830–42.

Mo, Q., S. F. Lu, et al. (2004). "DHEA and DHEA sulfate differentially regulate neural androgen receptor and its transcriptional activity." *Brain Res Mol Brain Res* 126(2): 165–72.

Moffat, S. D., and S. M. Resnick (2007). "Long-term measures of free testosterone predict regional cerebral blood flow patterns in elderly men." *Neurobiol Aging* 28(6): 914–20.

Mohr, C., A. C. Rowe, et al. (2010). "The influence of sex and empathy on putting oneself in the shoes of others." *Br J Psychol*, published online July 15, 2009.

Moisio, R. J. (2007). "Men in no-man's land: Proving manhood through compensatory consumption." Working paper presented at the ACRC in Memphis, TN.

Mondillon, L., P. M. Niedenthal, et al. (2007). "Imitation of in-group versus out-group members' facial expressions of anger: A test with a time perception task." *Soc Neurosci* 2(3–4): 223–37.

Mong, J. A., and D. W. Pfaff (2003). "Hormonal and genetic influences underlying arousal as it drives sex and aggression in animal and human brains." *Neurobiol Aging* 24, suppl. 1: S83–S88; discussion S91–S92.

Moons, W. G., and D. M. Mackie (2007). "Thinking straight while seeing red: The influence of anger on information processing." *Personality and Social Psychology Bulletin* 33(5): 706–20.

Moore, D. S., and S. P. Johnson (2008). "Mental rotation in human infants: A sex difference." *Psychol Sci* 19(11): 1063–66.

Moriguchi, Y., T. Ohnishi, et al. (2009). "The human mirror neuron system in a population with deficient self-awareness: An fMRI study in alexithymia." *Hum Brain Mapp* 30(7): 2063–76.

Morse, C. A., A. Buist, et al. (2000). "First-time parenthood: Influences on pre- and postnatal adjustment in fathers and mothers." *J Psychosom Obstet Gynaecol* 21(2): 109–20.

Motta, S. C., M. Goto, et al. (2009). "Dissecting the brain's fear system reveals the hypothalamus is critical for responding in subordinate conspecific intruders." *Proc Natl Acad Sci U S A* 106(12): 4870–75.

Moulier, V., H. Mouras, et al. (2006). "Neuroanatomical correlates of penile erection evoked by photographic stimuli in human males." *Neuroimage* 33(2): 689–99.

Mouras, H., S. Stoleru, et al. (2008). "Activation of mirror-neuron system by erotic video clips predicts degree of induced erection: An fMRI study." *Neuroimage* 42(3): 1142–50.

Muehlenhard, C. L., and S. K. Shippee (2009). "Men's and Women's Reports of Pretending Orgasm." *J Sex Res* 5: 1–16.

Mueller, S. C., D. Mandell, et al. (2009). "Early hyperandrogenism affects the development of hippocampal function: Preliminary evidence from a functional magnetic resonance imaging study of boys with familial male precocious puberty." *J Child Adolesc Psychopharmacol* 19(1): 41–50.

Muir, C. C., K. Treasurywala, et al. (2008). "Enzyme immunoassay of testosterone, 17beta-estradiol, and progesterone in perspiration and urine of preadolescents and young adults: Exceptional levels in men's axillary perspiration." *Horm Metab Res* 40(11): 819–26.

Mujica-Parodi, L. R., H. H. Strey, et al. (2009). "Chemosensory cues to conspecific emotional stress activate amygdala in humans." *PLoS One* 4(7): e6415.

Mulhall, J. P., R. King, et al. (2008a). "Evaluating the sexual experience in men: Validation of the Sexual Experience Questionnaire." *Journal of Sexual Medicine* 5(2): 365–76.

Mulhall, J., R. King, et al. (2008b). "Importance of and satisfaction with sex among men and women worldwide: Results of the Global Better Sex Survey." *Journal of Sexual Medicine* 5(4): 788–95.

Muller, M. N., F. W. Marlowe, et al. (2009). "Testosterone and paternal care in East African foragers and pastoralists." *Proc Biol Sci* 276(1655): 347–54.

Muller, M. N., and R. W. Wrangham (2004a). "Dominance, aggression and testosterone in wild chimpanzees: A test of the 'challenge hypothesis.' " *Animal Behaviour* 67(1): 113–23.

Muller, M. N., and R. W. Wrangham (2004b). "Dominance, cortisol and stress in wild chimpanzees *(Pan troglodytes schweinfurthii)*." *Behavioral Ecology and Sociobiology* 55(4): 332–40.

Mumme, D. L., A. Fernald, et al. (1996). "Infants' responses to facial and vocal emotional signals in a social referencing paradigm." *Child Dev* 67(6): 3219–37.

Munroe, R. L., and R. H. Munroe, eds. (1987). *The Couvade and Male Pregnancy Symptoms.* Berwyn, PA: Swets North America.

Munzert, J., B. Lorey, et al. (2009). "Cognitive motor processes: The role of motor imagery in the study of motor representations." *Brain Res Rev* 60(2): 306–26.

Murray, E. K., A. Hien, et al. (2009). "Epigenetic control of sexual differentiation of the bed nucleus of the stria terminalis." *Endocrinology* 150(9): 4241–47.

Murstein, B. I., and A. Tuerkheimer (1998). "Gender differences in love, sex, and motivation for sex." *Psychol Rep* 82(2): 435–50.

Mykletun, A., A. A. Dahl, et al. (2006). "Assessment of male sexual function by the Brief Sexual Function Inventory." *BJU Int* 97(2): 316–23.

Nakamura, Y., H. X. Gang, et al. (2009). "Adrenal changes associated with adrenarche." *Rev Endocr Metab Disord* 10(1): 19–26.

Nanova, P., L. Lyamova, et al. (2008). "Gender-specific development of auditory information processing in children: An ERP study." *Clin Neurophysiol* 119(9): 1992–2003.

Narring, F., S. M. Stronski Huwiler, et al. (2003). "Prevalence and dimensions of sexual orientation in Swiss adolescents: A cross-sectional survey of 16- to 20-year-old students." *Acta Paediatr* 92(2): 233–39.

National Council on Education (2009). "The condition of education," http://nces.ed.gov/programs/coe/.

Nealey-Moore, J. B., T. W. Smith, et al. (2007). "Cardiovascular reactivity during positive and negative marital interactions." *Journal of Behavioral Medicine* 30(6): 505–19.

Nelson, E. E., E. Leibenluft, et al. (2005). "The social re-orientation of adolescence: A neuroscience perspective on the process and its relation to psychopathology." *Psychological Medicine* 35(2): 163–74.

Nelson, R. J., and S. Chiavegatto (2001). "Molecular basis of aggression." *Trends Neurosci* 24(12): 713–19.

Neufang, S., K. Specht, et al. (2009). "Sex differences and the impact of steroid hormones on the developing human brain." *Cereb Cortex* 19(2): 464–73.

Neuhaus, A. H., C. Opgen-Rhein, et al. (2009). "Spatiotemporal mapping of sex differences during attentional processing." *Hum Brain Mapp* 30(9): 2997–3008.

Neumann, I. D. (2008a). "Brain oxytocin: A key regulator of emotional and social behaviours in both females and males." *J Neuroendocrinol* 20(6): 858–65.

Neumann, I. D., and R. Landgraf (2008b). "Advances in vasopressin and oxytocin—from genes to behaviour to disease: Preface." *Prog Brain Res* 170: xi–xiii.

Newcombe, N. S., and J. Huttenlocher, eds. (2006). *Development of Spatial Cognition*. Hoboken, NJ: John Wiley.

Newcombe, N. S., L. Mathason, et al., eds. (2002). *Maximization of Spatial Competence: More Important Than Finding the Cause of Sex Differences*. Westport, CT: Ablex Publishing.

Newman, M. L., C. J. Groom, et al. (2008). "Gender differences in language use: An analysis of 14,000 text samples." *Discourse Processes* 45(3): 211–36.

Newman, M. L., J. W. Pennebaker, et al. (2003). "Lying words: Predicting deception from linguistic styles." *Pers Soc Psychol Bull* 29(5): 665–75.

Newman, M. L., J. G. Sellers, et al. (2005). "Testosterone, cognition, and social status." *Horm Behav* 47(2): 205–11.

Niedenthal, P. M. (2007). "Embodying emotion." *Science* 316(5827): 1002–5.

Niederle, M. (2005). "Why do women shy away from competition? Do men compete too much?" *NBER, working paper*, July 2005.

Niel, L., A. H. Shah, et al. (2009). "Sexual differentiation of the spinal nucleus of the bulbocavernosus is not mediated solely by androgen receptors in muscle fibers." *Endocrinology* 150(7): 3207–13.

Nielsen, L., B. Knutson, et al. (2008). "Affect dynamics, affective forecasting, and aging." *Emotion* 8(3): 318–30.

Nikolova, G., and E. Vilain (2006). "Mechanisms of disease: Transcription factors in sex determination—relevance to human disorders of sex development." *Nat Clin Pract Endocrinol Metab* 2(4): 231–38.

Nummenmaa, L., J. Hirvonen, et al. (2008). "Is emotional contagion special? An fMRI study on neural systems for affective and cognitive empathy." *Neuroimage* 43(3): 571–80.

Nunez, J. L., H. A. Jurgens, et al. (2000). "Androgens reduce cell death in the developing rat visual cortex." *Brain Res Dev Brain Res* 125(1–2): 83–88.

Nuttall, R. L., M. B. Casey, et al., eds. (2005). *Spatial Ability as a Mediator of Gender Differences on Mathematics Tests: A Biological-Environmental Framework*. New York: Cambridge University Press.

O'Connor, D. B., J. Archer, et al. (2004). "Effects of testosterone on mood, aggression, and sexual behavior in young men: A double-blind, placebo-controlled, cross-over study." *J Clin Endocrinol Metab* 89(6): 2837–45.

O'Hair, D., and M. J. Cody (1987). "Gender and vocal stress differences during truthful and deceptive information sequences." *Human Relations* 40(1): 1–13.

O'Neill, C. T., L. J. Trainor, et al. (2001). "Infants' responsiveness to fathers' singing." *Music Perception* 18(4): 409–25.

Ochsner, K. N., R. D. Ray, et al. (2004). "For better or for worse: Neural systems supporting the cognitive down- and up-regulation of negative emotion." *Neuroimage* 23(2): 483–99.

Olson, C. K., L. A. Kutner, et al. (2007). "Factors correlated with violent video game use by adolescent boys and girls." *Journal of Adolescent Health* 41(1): 77–83.

Olsson, S. B., J. Barnard, et al. (2006). "Olfaction and identification of unrelated individuals: Examination of the mysteries of human odor recognition." *J Chem Ecol* 32(8): 1635–45.

Olweus, D., A. Mattsson, et al. (1988). "Circulating testosterone levels and aggression in adolescent males: A causal analysis." *Psychosom Med* 50(3): 261–72.

Olweus, D., A. Mattsson, et al. (1980). "Testosterone, aggression, physical, and personality dimensions in normal adolescent males." *Psychosom Med* 42(2): 253–69.

Ophir, A. G., J. O. Wolff, et al. (2008). "Variation in neural V1aR predicts sexual fidelity and space use among male prairie voles in semi-natural settings." *Proc Natl Acad Sci U S A* 105(4): 1249–54.

Ortigue, S., and F. Bianchi-Demicheli (2008). "The chronoarchitecture of human sexual desire: A high-density electrical mapping study." *Neuroimage* 43(2): 337–45.

Orzhekhovskaia, N. S. (2005). "[Sex dimorphism of neuron-glia correlations in the frontal areas of the human brain]." *Morfologiia* 127(1): 7–9.

Paick, J. S., J. H. Yang, et al. (2006). "The role of prolactin levels in the sexual activity of married men with erectile dysfunction." *BJU Int* 98(6): 1269–73.

Pak, T. R. (2008). "Sex differences in hormone receptors and behavior." In: J. B. Becker, K. Berkley, N. Geary, E. Hampson, J. P. Herman, and E. A. Young, eds., *Sex Differences in the Brain: From Genes to Behavior.* Oxford, UK: Oxford University Press.

Pak, T. R., W. C. Chung, et al. (2009). "Arginine vasopressin regulation in pre- and postpubertal male rats by the androgen metabolite 3beta-diol." *Am J Physiol Endocrinol Metab* 296(6): E1409–13.

Pancsofar, N., L. Vernon-Feagans, et al. (2008). "Family relationships during infancy and later mother and father vocabulary use with young children." *Early Child Res Q* 23(4): 493–503.

Panksepp, J. (2009). "Primary process affects and brain oxytocin." *Biol Psychiatry* 65(9): 725–27.

Paredes, R. G. (2009). "Evaluating the neurobiology of sexual reward." *ILAR J* 50(1): 15–27.

Parker Jr., C. R. (1999). "Dehydroepiandrosterone and dehydroepiandrosterone sulfate production in the human adrenal during development and aging." *Steroids* 64(9): 640–47.

Parker, J., and M. Burkley (2009). "Who's chasing whom? The impact of gender and relationship status on mate poaching." *Journal of Experimental Social Psychology* 45(4): 1016–19.

Parsons, T. D., P. Larson, et al. (2004). "Sex differences in mental rotation and spatial rotation in a virtual environment." *Neuropsychologia* 42(4): 555–62.

Pasley, K., T. G. Futris, et al. (2002). "Effects of commitment and psychological centrality on fathering." *Journal of Marriage and Family* 64(1): 130–38.

Pasterski, V., ed. (2008). *Disorders of sex development and atypical sex differentiation.* Hoboken, NJ: John Wiley.

Pasterski, V. L., M. E. Geffner, et al. (2005). "Prenatal hormones and postnatal socialization by parents as determinants of male-typical toy play in girls with congenital adrenal hyperplasia." *Child Dev* 76(1): 264–78.

Pasterski, V., P. Hindmarsh, et al. (2007). "Increased aggression and activity level in 3- to 11-year-old girls with congenital adrenal hyperplasia (CAH)." *Hormones and Behavior* 52(3): 368–74.

Paus, T., I. Nawaz-Khan, et al. (2010). "Sexual dimorphism in the adolescent brain: Role of testosterone and androgen receptor in global and local volumes of grey and white matter." *Horm Behav*, published online August 22, 2009.

Paus, T., A. Zijdenbos, et al. (1999). "Structural maturation of neural pathways in children and adolescents: In vivo study." *Science* 283(5409): 1908–11.

Pawlowski, B., L. G. Boothroyd, et al. (2008). "Is female attractiveness related to final reproductive success?" *Coll Antropol* 32(2): 457–60.

Payne, K., L. Thaler, et al. (2007). "Sensation and sexual arousal in circumcised and uncircumcised men." *J Sex Med* 4(3): 667–74.

Pecheux, M.-G., and F. Labrell, eds. (1994). *Parent-Infant Interactions and Early Cognitive Development*. Hillsdale, NJ: Lawrence Erlbaum Associates.

Peltola, M. J., J. M. Leppanen, et al. (2009). "Emergence of enhanced attention to fearful faces between 5 and 7 months of age." *Soc Cogn Affect Neurosci* 4(2): 134–42.

Penaloza, C., B. Estevez, et al. (2009). "Sex of the cell dictates its response: Differential gene expression and sensitivity to cell death inducing stress in male and female cells." *FASEB J* 23: 1869–79.

Pennebaker, J. W., C. J. Groom, et al. (2004). "Testosterone as a social inhibitor: Two case studies of the effect of testosterone treatment on language." *J Abnorm Psychol* 113(1): 172–75.

Peper, J. S., R. M. Brouwer, et al. (2009a). "Does having a twin brother make for a bigger brain?" *Eur J Endocrinol* 160(5): 739–46.

Peper, J. S., R. M. Brouwer, et al. (2009b). "Sex steroids and brain structure in pubertal boys and girls." *Psychoneuroendocrinology* 34(3): 332–42.

Perrin, J. S., G. Leonard, et al. (2009). "Sex differences in the growth of white matter during adolescence." *Neuroimage* 45(4): 1055–66.

Peters, M., W. Lehmann, et al. (2006). "Mental rotation test performance in four cross-cultural samples (n = 3367): Overall sex differences and the role of academic program in performance." *Cortex* 42(7): 1005–14.

Pfaff, D., ed. (2002). *Hormones, Brain and Behavior*. 5 vols.

Pfaff, D., E. Choleris, et al. (2005). "Genes for sex hormone receptors controlling mouse aggression." *Novartis Found Symp* 268: 78–89; discussion 89–99.

Pfeifer, J. H., C. L. Masten, et al. (2009). "Neural correlates of direct and reflected self-appraisals in adolescents and adults: When social perspective-taking informs self-perception." *Child Dev* 80(4): 1016–38.

Phelps, E. A. (2004). "Human emotion and memory: Interactions of the amygdala and hippocampal complex." *Curr Opin Neurobiol* 14(2): 198–202.

Piefke, M., and G. R. Fink (2005). "Recollections of one's own past: The effects of aging and gender on the neural mechanisms of episodic autobiographical memory." *Anat Embryol* (Berlin) 210(5–6): 497–512.

Piefke, M., P. H. Weiss, et al. (2005). "Gender differences in the functional neuroanatomy of emotional episodic autobiographical memory." *Hum Brain Mapp* 24(4): 313–24.

Pike, C. J., J. C. Carroll, et al. (2009). "Protective actions of sex steroid hormones in Alzheimer's disease." *Front Neuroendocrinol* 30(2): 239–58.

Pike, C. J., T. V. Nguyen, et al. (2008). "Androgen cell signaling pathways involved in neuroprotective actions." *Horm Behav* 53(5): 693–705.

Pillsworth, E. G., M. G. Haselton, et al. (2004). "Ovulatory shifts in female sexual desire." *J Sex Res* 41(1): 55–65.

Pinkerton, S. D., L. M. Bogart, et al. (2002). "Factors associated with masturbation in collegiate sample." *Journal of Psychology & Human Sexuality* 14(2–3): 103–21.

Pipitone, R. N., and G. G. Gallup, Jr. (2008). "Women's voice attractiveness varies across the menstrual cycle." *Evolution and Human Behavior* 29(4): 268–74.

Pittman, Q. J., and S. J. Spencer (2005). "Neurohypophysial peptides: Gatekeepers in the amygdala." *Trends Endocrinol Metab* 16(8): 343–44.

Plante, E., V. J. Schmithorst, et al. (2006). "Sex differences in the activation of language cortex during childhood." *Neuropsychologia* 44(7): 1210–21.

Ponseti, J., P. Kropp, et al. (2009). "Brain potentials related to the human penile erection." *Int J Impot Res* 21(5): 292–300.

Postma, A,, J. Winkel, et al. (1999). "Sex differences and menstrual cycle effects in human spatial memory." *Psychoneuroendocrinology* 24(2): 175–92.

Potegal, M., and J. Archer (2004). "Sex differences in childhood anger and aggression." *Child Adolesc Psychiatr Clin N Am* 13(3): vi–vii, 513–28.

Powell, F. D., L. D. Fields, et al. (2007). "Manhood, scholarship, perseverance, uplift, and elementary students: An example of school and community collaboration." *Urban Education* 42(4): 296–312.

Prehn-Kristensen, A., C. Wiesner, et al. (2009). "Induction of empathy by the smell of anxiety." *PloS One* 4(6): e5987.

Proverbio, A. M., R. Adorni, et al. (2009). "Sex differences in the brain response to affective scenes with or without humans." *Neuropsychologia* 47(12): 2374–88.

Proverbio, A. M., V. Brignone, et al. (2006a). "Gender differences in hemispheric asymmetry for face processing." *BMC Neurosci* 7: 44.

Proverbio, A. M., V. Brignone, et al. (2006b). "Gender and parental status affect the visual cortical response to infant facial expression." *Neuropsychologia* 44(14): 2987–99.

Proverbio, A. M., A. Zani, et al. (2008). "Neural markers of a greater female responsiveness to social stimuli." *BMC Neurosci* 9: 56.

Pruessner, J. C., F. Champagne, et al. (2004). "Dopamine release in response to a psychological stress in humans and its relationship to early life maternal care: A positron emission tomography study using [Â¹Â¹C] Raclopride." *Journal of Neuroscience* 24(11): 2825–31.

Puts, D. A., C. L. Jordan, et al. (2006a). "Defending the brain from estrogen." *Nat Neurosci* 9(2): 155–56.

Puts, D. A., C. L. Jordan, et al. (2006b). "O brother, where art thou? The fraternal birth-order effect on male sexual orientation." *Proc Natl Acad Sci U S A* 103(28): 10531–32.

Puts, D. A., M. A. McDaniel, et al. (2008). "Spatial ability and prenatal androgens: Meta-analyses of congenital adrenal hyperplasia and digit ratio (2D:4D) studies." *Arch Sex Behav* 37(1): 100–111.

Qian, S. Z., Y. Cheng Xu, et al. (2000). "Hormonal deficiency in elderly males." *Int J Androl* 23, suppl. 2: 1–3.

Quaiser-Pohl, C., and W. Lehmann (2002). "Girls' spatial abilities: Charting the contributions of experiences and attitudes in different academic groups." *Br J Educ Psychol* 72(2): 245–60.

Quigley, C. A. (2002). "Editorial: The postnatal gonadotropin and sex steroid surge—insights from the androgen insensitivity syndrome." *J Clin Endocrinol Metab* 87(1): 24–28.

Quinn, P. C., and L. S. Liben (2008). "A sex difference in mental rotation in young infants." *Psychol Sci* 19(11): 1067–70.

Raggenbass, M. (2008). "Overview of cellular electrophysiological actions of vasopressin." *Eur J Pharmacol* 583(2–3): 243–54.

Rahman, Q., D. Andersson, et al. (2005). "A specific sexual orientation–related difference in navigation strategy." *Behav Neurosci* 119(1): 311–16.

Rahman, Q., A. Cockburn, et al. (2008). "A comparative analysis of functional cerebral asymmetry in lesbian women, heterosexual women, and heterosexual men." *Arch Sex Behav* 37(4): 566–71.

Rainey, W. E., and Y. Nakamura (2008). "Regulation of the adrenal androgen biosynthesis." *J Steroid Biochem Mol Biol* 108(3–5): 281–86.

Rajender, S., G. Pandu, et al. (2008). "Reduced CAG repeats length in androgen receptor gene is associated with violent criminal behavior." *Int J Legal Med* 122(5): 367–72.

Rajpert-De Meyts, E., N. Jorgensen, et al. (1999). "Expression of anti-Müllerian hormone during normal and pathological gonadal development: Association with differentiation of Sertoli and granulosa cells." *J Clin Endocrinol Metab* 84(10): 3836–44.

Reber, S. O., and I. D. Neumann (2008). "Defensive behavioral strategies and enhanced state anxiety during chronic subordinate colony housing are accompanied by reduced hypothalamic vasopressin, but not oxytocin, expression." *Ann N Y Acad Sci* 1148: 184–95.

Reburn, C. J., and K. E. Wynne-Edwards (1999). "Hormonal changes in males of a naturally biparental and a uniparental mammal." *Horm Behav* 35(2): 163–76.

Redeker, G. (2008). "Gendered discourse practices in instant messaging." Talk given at University of Groningen, Nov. 15, 2008.

Redoute, J., S. Stoleru, et al. (2005). "Brain processing of visual sexual stimuli in treated and untreated hypogonadal patients." *Psychoneuroendocrinology* 30(5): 461–82.

Redoute, J., S. Stoleru, et al. (2000). "Brain processing of visual sexual stimuli in human males." *Hum Brain Mapp* 11(3): 162–77.

Rehman, K. S., and B. R. Carr (2004). "Sex differences in adrenal androgens." *Semin Reprod Med* 22(4): 349–60.

Reinius, B., P. Saetre, et al. (2008). "An evolutionarily conserved sexual signature in the primate brain." *PLoS Genet* 4(6): e1000100.

Resnick, S. M. (2008). "Sex differences in brain aging." In: J. B. Becker, K. Berkley, N. Geary, E. Hampson, J. P. Herman, and E. A. Young, eds., *Sex Differences in the Brain: From Genes to Behavior.* Oxford, UK: Oxford University Press.

Revicki, D., K. Howard, et al. (2008). "Characterizing the burden of premature ejaculation from a patient and partner perspective: A multi-country qualitative analysis." *Health Qual Life Outcomes* 6: 33.

Richters, J., R. Visser, et al. (2006). "Sexual practices at last heterosexual encounter and occurrence of orgasm in a national survey." *J Sex Res* 43(3): 217–26.

Rilling, J. K., J. T. Winslow, et al. (2004). "The neural correlates of mate competition in dominant male rhesus macaques." *Biol Psychiatry* 56(5): 364–75.

Roberto, K. A., K. R. Allen, et al. (2001). "Grandfathers' perceptions and expectations of relationships with their adult grandchildren." *Journal of Family Issues* 22(4): 407–26.

Roberts, S. C., L. M. Gosling, et al. (2005). "Body odor similarity in noncohabiting twins." *Chem Senses* 30(8): 651–56.

Roberts, S. C., and A. C. Little (2008). "Good genes, complementary genes and human mate preferences." *Genetica* 134(1): 31–43.

Robinson, G. E., R. D. Fernald, et al. (2008). "Genes and social behavior." *Science* 322(5903): 896–900.

Roenneberg, T., T. Kuehnle, et al. (2004). "A marker for the end of adolescence." *Curr Biol* 14(24): R1038–39.

Roese, N. J., G. L. Pennington, et al. (2006). "Sex differences in regret: All for love or some for lust?" *Pers Soc Psychol Bull* 32(6): 770–80.

Romeo, R. D., S. L. Diedrich, et al. (2000). "Effects of gonadal steroids during pubertal development on androgen and estrogen receptor-alpha immunoreactivity in the hypothalamus and amygdala." *J Neurobiol* 44(3): 361–68.

Roney, J. R., K. N. Hanson, et al. (2006). "Reading men's faces: Women's mate attractiveness judgments track men's testosterone and interest in infants." *Proc Biol Sci* 273(1598): 2169–75.

Roney, J. R., A. W. Lukaszewski, et al. (2007). "Rapid endocrine responses of young men to social interactions with young women." *Horm Behav* 52(3): 326–33.

Roney, J. R., and Z. L. Simmons (2008). "Women's estradiol predicts preference for facial cues of men's testosterone." *Horm Behav* 53(1): 14–19.

Roopnarine, J. L., H. N. Fouts, et al. (2005). "Mothers' and fathers' behaviors toward their 3- to 4-month-old infants in lower, middle, and upper socioeconomic African American families." *Developmental Psychology* 41(5): 723–32.

Rosario, E. R., L. Chang, et al. (2009). "Brain levels of sex steroid hormones in men and women during normal aging and in Alzheimer's disease." *Neurobiol Aging*, pubilshed online May 9, 2009.

Rosario, E. R., L. Chang, et al. (2004). "Age-related testosterone depletion and the development of Alzheimer disease." *JAMA* 292(12): 1431–32.

Rosario, E. R., and C. J. Pike (2008). "Androgen regulation of beta-amyloid protein and the risk of Alzheimer's disease." *Brain Res Rev* 57(2): 444–53.

Rose, A. B., D. P. Merke, et al. (2004). "Effects of hormones and sex chromosomes on stress-influenced regions of the developing pediatric brain." *Ann N Y Acad Sci* 1032: 231–33.

Rose, A. J., and K. D. Rudolph (2006). "A review of sex differences in peer relationship processes: Potential trade-offs for the emotional and behavioral development of girls and boys." *Psychol Bull* 132(1): 98–131.

Rosen, R., E. Janssen, et al. (2006). "Psychological and interpersonal correlates in men with erectile dysfunction and their partners: A pilot study of treatment outcome with sildenafil." *J Sex Marital Ther* 32(3): 215–34.

Rosen, W. D., L. B. Adamson, et al. (1992). "An experimental investigation of infant social referencing: Mothers' messages and gender differences." *Developmental Psychology* 28(6): 1172–78.

Rosip, J.C.H., and A. Judith (2004). "Knowledge of nonverbal cues, gender, and nonverbal decoding accuracy." *Journal of Nonverbal Behavior, special, Interpersonal Sensitivity*, pt. 2. 28(4): 267–86.

Rowe, R., B. Maughan, et al. (2004). "Testosterone, antisocial behavior, and social dominance in boys: Pubertal development and biosocial interaction." *Biol Psychiatry* 55(5): 546–52.

Rubin, R. H., and D. W. Pfaff. (2009). *Hormone/Behavior Relations of Clinical Importance: Endocrine Systems Interacting with Brain and Behavior.* London: Cambridge University Press.

Rubinow, D. R., C. A. Roca, et al. (2005). "Testosterone suppression of CRH-stimulated cortisol in men." *Neuropsychopharmacology* 30(10): 1906–12.

Ruytjens, L., J. R. Georgiadis, et al. (2007). "Functional sex differences in human primary auditory cortex." *Eur J Nucl Med Mol Imaging* 34(12): 2073–81.

Rymarczyk, K., and A. Grabowska (2007). "Sex differences in brain control of prosody." *Neuropsychologia* 45(5): 921–30.

Saad, F., A. Kamischke, et al. (2007). "More than eight years' hands-on experience with the novel long-acting parenteral testosterone undecanoate." *Asian J Androl* 9(3): 291–97.

Sadeghi-Nejad, H., and R. Watson (2008). "Premature ejaculation: Current medical treatment and new directions (CME)." *J Sex Med* 5(5): 1037–50; quiz, 1051–52.

Sakalli-Ugurlu, N. (2003). "How do romantic relationship satisfaction, gender stereotypes, and gender relate to future time orientation in romantic relationships?" *J Psychol* 137(3): 294–303.

Sakuma, Y. (2009). "Gonadal steroid action and brain sex differentiation in the rat." *J Neuroendocrinol* 21(4): 410–14.

Sallet, J., and M. F. Rushworth (2009). "Should I stay or should I go: Genetic bases for uncertainty-driven exploration." *Nat Neurosci* 12(8): 963–65.

Salvador, A. (2005). "Coping with competitive situations in humans." *Neuroscience & Biobehavioral Reviews* 29: 195–205.

Salvador, A., V. Simon, F. Suay, and L. Llorens (1987). "Testosterone and cortisol responses to competitive fighting: A pilot study." *Aggressive Behavior* 13: 9–13.

Salvador, A., F. Suay, E. González-Bono, and M. A. Serrano (2003). "Anticipatory cortisol, testosterone and psychological responses to judo competition in young men." *Psychoneuroendocrinology* 28:364–75.

Samanez-Larkin, G. R., S. E. Gibbs, et al. (2007). "Anticipation of monetary gain but not loss in healthy older adults." *Nat Neurosci* 10(6): 787–91.

Sanchez, D. T., and A. K. Kiefer (2007). "Body concerns in and out of the bedroom: Implications for sexual pleasure and problems." *Arch Sex Behav* 36(6): 808–20.

Sanchez-Martin, J. R., E. Fano, et al. (2000). "Relating testosterone levels and free play social behavior in male and female preschool children." *Psychoneuroendocrinology* 25(8): 773–83.

Sanchez Rodriguez, S. M., F. Pelaez del Hierro, et al. (2008). "Body weight increase in expectant males and helpers of cotton-top tamarin (*Saguinus oedipus*): A symptom of the couvade syndrome?" *Psicothema* 20(4): 825–29.

Sand, M. S., W. Fisher, et al. (2008). "Erectile dysfunction and constructs of masculinity and quality of life in the multinational Men's Attitudes to Life Events and Sexuality (MALES) study." *J Sex Med* 5(3): 583–94.

Santos, P. S., J. A. Schinemann, et al. (2005). "New evidence that the MHC influences odor perception in humans: A study with 58 southern Brazilian students." *Horm Behav* 47(4): 384–88.

Sapolsky, R. M. (2005). "The influence of social hierarchy on primate health." *Science* 308(5722): 648–52.

Sapolsky, R. M. (1986). "Stress-induced elevation of testosterone concentration in high ranking baboons: Role of catecholamines." *Endocrinology* 118(4): 1630–35.

Sapolsky, R. M., and M. J. Meaney (1986). "Maturation of the adrenocortical stress response: Neuroendocrine control mechanisms and the stress hyporesponsive period." *Brain Res* 396(1): 64–76.

Sapolsky, R. M., J. H. Vogelman, et al. (1993). "Senescent decline in serum dehydroepiandrosterone sulfate concentrations in a population of wild baboons." *J Gerontol* 48(5): B196–B200.

Sarkadi, A., R. Kristiansson, et al. (2008). "Fathers' involvement and children's developmental outcomes: A systematic review of longitudinal studies." *Acta Paediatr* 97(2): 153–58.

Sato, S. M., K. M. Schulz, et al. (2008). "Adolescents and androgens, receptors and rewards." *Horm Behav* 53(5): 647–58.

Savic, I. (2001a). "Processing of odorous signals in humans." *Brain Res Bull* 54(3): 307–12.

Savic, I., H. Berglund, et al. (2005). "Brain response to putative pheromones in homosexual men." *Proc Natl Acad Sci U S A* 102(20): 7356–61.

Savic, I., H. Berglund, et al. (2001b). "Smelling of odorous sex hormone–like compounds causes sex-differentiated hypothalamic activations in humans." *Neuron* 31(4): 661–68.

Savic, I., E. Heden-Blomqvist, et al. (2009). "Pheromone signal transduction in humans: What can be learned from olfactory loss." *Hum Brain Mapp* 30(9): 3057–65.

Savic, I., and P. Lindstrom (2008). "PET and MRI show differences in cerebral asymmetry and functional connectivity between homo- and heterosexual subjects." *Proc Natl Acad Sci U S A* 105(27): 9403–8.

Savulescu, J., and A. Sandberg (2008). "Neuroenhancement of love and marriage: The chemicals between us." *Neuroethics* 1(1): 31–44.

Saxton, T. K., A. Lyndon, et al. (2008). "Evidence that androstadienone, a putative human chemosignal, modulates women's attributions of men's attractiveness." *Horm Behav* 54(5): 597–601.

Schacht, A., and W. Sommer (2009). "Emotions in word and face processing: Early and late cortical responses." *Brain Cogn* 69(3): 538–50.

Schirmer, A., N. Escoffier, et al. (2008). "What grabs his attention but not hers? Estrogen correlates with neurophysiological measures of vocal change detection." *Psychoneuroendocrinology* 33(6): 718–27.

Schirmer, A., and S. A. Kotz (2003). "ERP evidence for a sex-specific Stroop effect in emotional speech." *J Cogn Neurosci* 15(8): 1135–48.

Schirmer, A., S. A. Kotz, et al. (2002). "Sex differentiates the role of emotional prosody during word processing." *Brain Res Cogn Brain Res* 14(2): 228–33.

Schmidt, J. A., J. M. Oatley, et al. (2009). "Female mice delay reproductive aging in males." *Biol Reprod* 80(5): 1009–14.

Schmithorst, V. J., S. K. Holland, et al. (2008). "Developmental differences in white matter architecture between boys and girls." *Hum Brain Mapp* 29(6): 696–710.

Schmitt, D. P. (2002). "A meta-analysis of sex differences in romantic attraction: Do rating contexts moderate tactic effectiveness judgments?" *Br J Soc Psychol* 41(3): 387–402.

Schmitt, D. P., L. Alcalay, et al. (2004). "Patterns and universals of mate poaching across 53 nations: The effects of sex, culture, and personality on romantically attracting another person's partner." *J Pers Soc Psychol* 86(4): 560–84.

Schmitt, D. P., and D. M. Buss (2001). "Human mate poaching: Tactics and temptations for infiltrating existing mateships." *Journal of Personality and Social Psychology* 80(6): 894–917.

Schmitt, D. P., and D. M. Buss (1996). "Strategic self-promotion and competitor derogation: Sex and context effects on the perceived effectiveness of mate attraction tactics." *J Pers Soc Psychol* 70(6): 1185–1204.

Schmitt, D. P., A. Couden, et al. (2001). "The effects of sex and temporal context on feelings of romantic desire: An experimental evaluation of sexual strategies theory." *Pers Soc Psychol Bull* 27(7): 833–47.

Schmitt, D. P., and T. K. Shackelford (2003). "Nifty ways to leave your lover: The tactics people use to entice and disguise the process of human mate poaching." *Pers Soc Psychol Bull* 29(8): 1018–35.

Schmitt, D. P., T. K. Shackelford, et al. (2001). "The desire for sexual variety as a key to understanding basic human mating strategies." *Personal Relationships, Special Issue: Evolutionary approaches to relationships* 8(4): 425–55.

Schmitt, M., M. Kliegel, et al. (2007). "Marital interaction in middle and old age: A predictor of marital satisfaction?" *Int J Aging Hum Dev* 65(4): 283–300.

Schober, J. M., and D. Pfaff (2007). "The neurophysiology of sexual arousal." *Best Pract Res Clin Endocrinol Metab* 21(3): 445–61.

Schoning, S., A. Engelien, et al. (2010). "Neuroimaging differences in spatial cognition between men and male-to-female transsexuals before and during hormone therapy." *J Sex Med*, published online September 14, 2009.

Schoppe-Sullivan, S. J., G. L. Brown, et al. (2008). "Maternal gatekeeping, coparenting quality, and fathering behavior in families with infants." *J Fam Psychol* 22(3): 389–98.

Schoppe-Sullivan, S. J., A. H. Weldon, et al. (2010). "Coparenting behavior moderates longitudinal relations between effortful control and preschool children's externalizing behavior." *J Child Psychol Psychiatry* 50(6): 698–706.

Schulte-Ruther, M., H. J. Markowitsch, et al. (2008). "Gender differences in brain networks supporting empathy." *Neuroimage* 42(1): 393–403.

Schultheiss, O. C., A. Dargel, et al. (2003). "Implicit motives and gonadal steroid hormones: Effects of menstrual cycle phase, oral contraceptive use, and relationship status." *Horm Behav* 43(2): 293–301.

Schultheiss, O. C., and W. Rohde (2002). "Implicit power motivation predicts men's testosterone changes and implicit learning in a contest situation." *Horm Behav* 41(2): 195–202.

Schultheiss, O. C., M. M. Wirth, et al. (2008). "Exploring the motivational brain: Effects of implicit power motivation on brain activation in response to facial expressions of emotion." *Soc Cogn Affect Neurosci* 3(4): 333–43.

Schultheiss, O. C., M. M. Wirth, et al. (2005). "Effects of implicit power motivation on men's and women's implicit learning and testosterone changes after social victory or defeat." *J Pers Soc Psychol* 88(1): 174–88.

Schulz, K. M., T. A. Menard, et al. (2006a). "Testicular hormone exposure during adolescence organizes flank-marking behavior and vasopressin receptor binding in the lateral septum." *Horm Behav* 50(3): 477–83.

Schulz, K. M., H. A. Molenda-Figueira, et al. (2009a). "Back to the future: The organizational-activational hypothesis adapted to puberty and adolescence." *Horm Behav* 55(5): 597–604.

Schulz, K. M., and C. L. Sisk (2006b). "Pubertal hormones, the adolescent brain, and the maturation of social behaviors: Lessons from the Syrian hamster." *Mol Cell Endocrinol* 254–55: 120–26.

Schulz, K. M., J. L. Zehr, et al. (2009b). "Testosterone programs adult social behavior before and during, but not after, adolescence." *Endocrinology* 150(8): 3690–98.

Schwarz, J. M., S. L. Liang, et al. (2008). "Estradiol induces hypothalamic dendritic spines by enhancing glutamate release: A mechanism for organizational sex differences." *Neuron* 58(4): 584–98.

Schwarz, J. M., and M. M. McCarthy (2008). "Steroid-induced sexual differentiation of the developing brain: Multiple pathways, one goal." *J Neurochem* 105(5): 1561–72.

Schweinsburg, A. D., B. J. Nagel, et al. (2005). "fMRI reveals alteration of spatial working memory networks across adolescence." *J Int Neuropsychol Soc* 11(5): 631–44.

Scordalakes, E. M., and E. F. Rissman (2004). "Aggression and arginine vasopressin immunoreactivity regulation by androgen receptor and estrogen receptor alpha." *Genes Brain Behav* 3(1): 20–26.

Sear, R., and R. Mace (2008). "Who keeps children alive? A review of the effects of kin on child survival." *Evolution and Human Behavior* 29(1): 1–18.

Seidman, S. N., and S. P. Roose (2006). "The sexual effects of testosterone replacement in depressed men: Randomized, placebo-controlled clinical trial." *J Sex Marital Ther* 32(3): 267–73.

Seifritz, E., F. Esposito, et al. (2003). "Differential sex-independent amygdala response to infant crying and laughing in parents versus nonparents." *Biol Psychiatry* 54(12): 1367–75.

Sell, A., L. Cosmides, et al. (2009). "Human adaptations for the visual assessment of strength and fighting ability from the body and face." *Proc Biol Sci* 276(1656): 575–84.

Seney, M. L., and N. G. Forger (2009). "Sexual differentiation of the nervous system: Where the action is." *Endocrinology* 150(7): 2991–93.

Sengezer, M., S. Ozturk, et al. (2002). "Accurate method for determining functional penile length in Turkish young men." *Ann Plast Surg* 48(4): 381–85.

Sergeant, M. J., T. E. Dickins, et al. (2007). "Women's hedonic ratings of body odor of heterosexual and homosexual men." *Arch Sex Behav* 36(3): 395–401.

Servin, A., A. Nordenstrom, et al. (2003). "Prenatal androgens and gender-typed behavior: A study of girls with mild and severe forms of congenital adrenal hyperplasia." *Dev Psychol* 39(3): 440–50.

Shackelford, T. K., and A. T. Goetz, eds. (2006a). *Predicting Violence Against Women from Men's Mate-Retention Behaviors.* New York: Cambridge University Press.

Shackelford, T. K., A. T. Goetz, et al. (2006b). "Mate guarding and frequent in-pair copulation in humans: Concurrent or compensatory anti-cuckoldry tactics?" *Human Nature, Special Issue: Human sperm competition* 17(3): 239–52.

Shackelford, T. K., A. T. Goetz, et al. (2005a). "Mate retention in marriage: Further evidence of the reliability of the Mate Retention Inventory." *Personality and Individual Differences* 39(2): 415–25.

Shackelford, T. K., A. T. Goetz, et al. (2005b). "When we hurt the ones we love: Predicting violence against women from men's mate retention." *Personal Relationships* 12(4): 447–63.

Shackelford, T. K., A. T. Goetz, et al. (2004). "Sex differences in sexual psychology produce sex-similar preferences for a short-term mate." *Archives of Sexual Behavior* 33(4): 405–12.

Shackelford, T. K., D. P. Schmitt, et al. (2005c). "Mate preferences of married persons in the newlywed year and three years later." *Cognition & Emotion* 19(8): 1262–70.

Shackelford, T. K., D. P. Schmitt, et al. (2005d). "Universal dimensions of human mate preferences." *Personality and Individual Differences* 39(2): 447–58.

Shafik, A. (1998). "The mechanism of ejaculation: The glans-vasal and urethromuscular reflexes." *Arch Androl* 41(2): 71–78.

Shafik, A., A. A. Shafik, et al. (2009). "Electromyographic study of ejaculatory mechanism." *Int J Androl* 32(3): 212–17.

Shafik, A., I. A. Shafik, et al. (2007). "The effect of external urethral sphincter contraction on the cavernosus muscles and its role in the sexual act." *Int Urol Nephrol* 39(2): 541–46.

Shah, J., and N. Christopher (2002). "Can shoe size predict penile length?" *BJU Int* 90(6): 586–87.

Shah, N. M., and S. M. Breedlove (2007). "Behavioural neurobiology: Females can also be from Mars." *Nature* 448(7157): 999–1000.

Shah, N. M., D. J. Pisapia, et al. (2004). "Visualizing sexual dimorphism in the brain." *Neuron* 43(3): 313–19.

Shamay-Tsoory, S. G., J. Aharon-Peretz, et al. (2009). "Two systems for empathy: A double dissociation between emotional and cognitive empathy in inferior frontal gyrus versus ventromedial prefrontal lesions." *Brain* 132(3): 617–27.

Shane, M. S., M. C. Stevens, et al. (2009). "Double dissociation between perspective-taking and empathic-concern as predictors of hemodynamic response to another's mistakes." *Soc Cogn Affect Neurosci* 4(2): 111–18.

Sharma, V., and P. Perros (2009). "The management of hypogonadism in aging male patients." *Postgrad Med* 121(1): 113–21.

Shaw, P., D. Greenstein, et al. (2006). "Intellectual ability and cortical development in children and adolescents." *Nature* 440(7084): 676–79.

Shaw, P., N. J. Kabani, et al. (2008). "Neurodevelopmental trajectories of the human cerebral cortex." *J Neurosci* 28(14): 3586–94.

Shaywitz, B. A., S. E. Shaywitz, et al. (1995). "Sex differences in the functional organization of the brain for language." *Nature* 373(6515): 607–9.

Sheldon, A., and L. Rohleder, eds. (1996). *Sharing the Same World, Telling Different Stories: Gender Differences in Co-constructed Pretend Narratives.* Hillsdale, NJ: Lawrence Erlbaum Associates.

Shepard, K. N., V. Michopoulos, et al. (2010). "Genetic, epigenetic and environmental impact on sex differences in social behavior." *Physiol Behav* 97(2): 157–70.

Shoup, M. L., and G. G. Gallup Jr. (2008). "Men's faces convey information about their bodies and their behavior: What you see is what you get." *Evolutionary Psychology* 6(3): 469–79.

Shulman, S., L. C. Mayes, et al. (2008). "Romantic attraction and conflict negotiation among late adolescent and early adult romantic couples." *J Adolesc* 31(6): 729–45.

Shultz, S., and R. I. Dunbar (2007). "The evolution of the social brain: Anthropoid primates contrast with other vertebrates." *Proc Biol Sci* 274(1624): 2429–36.

Shuster, S. (2007). "Sex, aggression, and humour: Responses to unicycling." *BMJ* 335(7633): 1320–22.

Siakaluk, P. D., P. M. Pexman, et al. (2008). "Evidence for the activation of sensorimotor information during visual word recognition: The body-object interaction effect." *Cognition* 106(1): 433–43.

Siegel, L. A., and R. M. Siegel (2007). "Sexual changes in the aging male." In: A. F. Owens and M. S. Tepper, eds., *Sexual Health,* vol 2: *Physical Foundations.* Westport, CT: Praeger/Greenwood, 223–55.

Siegel, M., T. H. Donner, et al. (2008). "Neuronal synchronization along the dorsal visual pathway reflects the focus of spatial attention." *Neuron* 60(4): 709–19.

Silk, J. B., J. C. Beehner, et al. (2009). "The benefits of social capital: Close social bonds among female baboons enhance offspring survival." *Proc Biol Sci* 276(1670): 3099–3104.

Silverstein, M., and A. Marenco (2001). "How Americans enact the grandparent role across the family life course." *Journal of Family Issues* 22(4): 493–522.

Simon, N. G., A. Cologer-Clifford, et al. (1998). "Testosterone and its metabolites modulate 5HT1A and 5HT1B agonist effects on intermale aggression." *Neurosci Biobehav Rev* 23(2): 325–36.

Simon, N. G., Q. Mo, et al. (2006). "Hormonal pathways regulating intermale and interfemale aggression." *Int Rev Neurobiol* 73: 99–123.

Singer, T., and C. Lamm (2009). "The social neuroscience of empathy." *Ann N Y Acad Sci* 1156: 81–96.

Singh, D. (2002). "Female mate value at a glance: Relationship of waist-to-hip ratio to health, fecundity and attractiveness." *Neuro Endocrinol Lett* 23 Suppl 4: 81–91.

Smiler, A. P. (2008). "'I wanted to get to know her better': Adolescent boys' dating motives, masculinity ideology, and sexual behavior." *J Adolesc* 31(1): 17–32.

Smith, L. J., J. P. Mulhall, et al. (2007). "Sex after seventy: A pilot study of sexual function in older persons." *J Sex Med* 4(5): 1247–53.

Smith, T. W., B. N. Uchino, et al. (2009). "Conflict and collaboration in middle-aged and older couples, pt. 2: Cardiovascular reactivity during marital interaction." *Psychol Aging* 24(2): 274–86.

Snowdon, C. T., T. E. Ziegler, et al. (2006). "Social odours, sexual arousal and pairbonding in primates." *Philos Trans R Soc Lond B Biol Sci* 361(1476): 2079–89.

Snyder, J. K., L. A. Kirkpatrick, et al. (2008). "The dominance dilemma: Do women really prefer dominant mates?" *Personal Relationships* 15(4): 425–44.

Snyder, P. J. (2008a). "Decreasing testosterone with increasing age: More factors, more questions." *J Clin Endocrinol Metab* 93(7): 2477–78.

Snyder, P. J. (2008b). "Might testosterone actually reduce mortality?" *J Clin Endocrinol Metab* 93(1): 32–33.

Sokhi, D. S., M. D. Hunter, et al. (2005). "Male and female voices activate distinct regions in the male brain." *Neuroimage* 27(3): 572–78.

Soldin, O. P., E. G. Hoffman, et al. (2005). "Pediatric reference intervals for FSH, LH, estradiol, T3, free T3, cortisol, and growth hormone on the DPC IMMULITE 1000." *Clin Chim Acta* 355(1–2): 205–10.

Soma, K. K., M. A. Scotti, et al. (2008). "Novel mechanisms for neuroendocrine regulation of aggression." *Front Neuroendocrinol* 29(4): 476–89.

Sonnby-Borgström, M., P. Jonsson, et al. (2008). "Gender differences in facial imitation and verbally reported emotional contagion from spontaneous to emotionally regulated processing levels." *Scand J Psychol* 49(2): 111–22.

Soulliere, D. M. (2006). "Wrestling with masculinity: Messages about manhood in the WWE." *Sex Roles* 55(1–2): 1–11.

Spear, L. P. (2004). "Adolescent brain development and animal models." *Ann N Y Acad Sci* 1021: 23–26.

Spelke, E. (2005). "The science of gender and science." *Edge*, May 15, 2005.

Spelke, E. S. (2005). "Sex differences in intrinsic aptitude for mathematics and science? A critical review." *Am Psychol* 60(9): 950–58.

Spence, I. (2009). "Women match men when learning a spatial skill." *Journal of Experimental Psychology*, special issue, *Learning, memory, and cognition* 35(4): pp. 1097–1103.

Spors, H., and N. Sobel (2007). "Male behavior by knockout." *Neuron* 55(5): 689–93.

Sprecher, S. (2002). "Sexual satisfaction in premarital relationships: Associations with satisfaction, love, commitment, and stability." *J Sex Res* 39(3): 190–96.

Srinivas-Shankar, U., and D. Sharma (2009a). "Testosterone treatment in elderly men." *Adv Ther* 26(1): 25–39.

Srinivas-Shankar, U., and F. C. Wu (2009b). "Frailty and muscle function: Role for testosterone?" *Front Horm Res* 37: 133–49.

St. Jacques, P. L., B. Bessette-Symons, et al. (2009). "Functional neuroimaging studies of aging and emotion: Fronto-amygdalar differences during emotional perception and episodic memory." *J Int Neuropsychol Soc* 15(6): 819–25.

Stanton, S. J., J. C. Beehner, et al. (2009a). "Dominance, politics, and physiology: Voters' testosterone changes on the night of the 2008 United States presidential election." *PloS One* 4(10): e7543.

Stanton, S. J., O. C. Schultheiss (2007). "Basal and dynamic relationships between implicit power motivation and estradiol in women." *Horm Behav* 52(5): 571–80.

Stanton, S. J., M. M. Wirth, et al. (2009b). "Endogenous testosterone levels are associated with amygdala and ventromedial prefrontal cortex responses to anger faces in men but not women." *Biol Psychol* 81(2): 118–22.

Starratt, V. G., D. Popp, et al. (2008). "Not all men are sexually coercive: A preliminary investigation of the moderating effect of mate desirability on the relationship between female infidelity and male sexual coercion." *Personality and Individual Differences* 45(1): 10–14.

Starratt, V. G., T. K. Shackelford, et al. (2007). "Male mate retention behaviors vary with risk of partner infidelity and sperm competition." *Acta Psychologica Sinica*, special issue, *Evolutionary Psychology* 39(3): 523–27.

Steers, W. D. (2000). "Neural pathways and central sites involved in penile erection: Neuroanatomy and clinical implications." *Neurosci Biobehav Rev* 24(5): 507–16.

Stein, D. J., J. van Honk, et al. (2007). "Opioids: From physical pain to the pain of social isolation." *CNS Spectr* 12(9): 669–74.

Steinberg, L. (2007). "Risk taking in adolescence: New perspectives from brain and behavioral science." *Current Directions in Psychological Science* 16(2): 55–59.

Steinberg, L., ed. (2004a). *Risk Taking in Adolescence: What Changes, and Why?* New York: New York Academy of Sciences.

Steinberg, L., and R. M. Lerner (2004b). "The scientific study of adolescence: A brief history." *Journal of Early Adolescence*, special issue, *Adolescence: The Legacy of Hershel and Ellen Thornburg* 24(1): 45–54.

Steiner, M., and E. A. Young (2008). "Hormones and mood." In: J. B. Becker, K. Berkley, N. Geary, E. Hampson, J. P. Herman, and E. A. Young, eds., *Sex Differences in the Brain: From Genes to Behavior.* Oxford, UK: Oxford University Press.

Stoleru, S., J. Redoute, et al. (2003). "Brain processing of visual sexual stimuli in men with hypoactive sexual desire disorder." *Psychiatry Res* 124(2): 67–86.

Storey, A. E., C. J. Walsh, et al. (2000). "Hormonal correlates of paternal responsiveness in new and expectant fathers." *Evol Hum Behav* 21(2): 79–95.

Strathearn, L., P. Fonagy, et al. (2009). "Adult attachment predicts maternal brain and oxytocin response to infant cues." *Neuropsychopharmacology* 34(13): 2655–66.

Striano, T., V. M. Reid, et al. (2006). "Neural mechanisms of joint attention in infancy." *Eur J Neurosci* 23(10): 2819–23.

Stroud, L. R., G. D. Papandonatos, et al. (2004). "Sex differences in the effects of pubertal development on responses to a corticotropin-releasing hormone challenge: The Pittsburgh psychobiologic studies." *Ann N Y Acad Sci* 1021: 348–51.

Stroud, L. R., P. Salovey, et al. (2002). "Sex differences in stress responses: Social rejection versus achievement stress." *Biol Psychiatry* 52(4): 318–27.

Suay, F., A. Salvador, E. González-Bono, C. Sanchis, M. Martínez, and S. Martínez-Sanchis (1999). "Effects of competition and its outcome on serum testosterone, cortisol and prolactin." *Psychoneuroendocrinology* 24: 551–66.

Summers, C. H., G. L. Forster, et al. (2005). "Dynamics and mechanics of social rank reversal." *J Comp Physiol A Neuroethol Sens Neural Behav Physiol* 191(3): 241–52.

Swaab, D. F. (2008). "Sexual orientation and its basis in brain structure and function." *Proc Natl Acad Sci U S A* 105(30): 10273–74.

Swaab, D. F. (2004). "The human hypothalamus. Basic and clinical aspects, part 2: Neuropathology of the hypothalamus and adjacent brain structures." In: F. Boller and D. F. Swaab, eds., *Handbook of Clinical Neurology.* Amsterdam: Elsevier.

Swaab, D. F., and E. Fliers (1985). "A sexually dimorphic nucleus in the human brain." *Science* 228(4703): 1112–15.

Swaab, D. F., and A. Garcia-Falgueras (2009). "Sexual differentiation of the human brain in relation to gender identity and sexual orientation." *Funct Neurol* 24(1): 17–28.

Swaab, D. F., L. J. Gooren, et al. (1995). "Brain research, gender and sexual orientation." *J Homosex* 28(3–4): 283–301.

Swaab, D. F., and M. A. Hofman (1990). "An enlarged suprachiasmatic nucleus in homosexual men." *Brain Res* 537(1–2): 141–48.

Swain, J. E. (2008). "Baby stimuli and the parent brain: Functional neuroimaging of the neural substrates of parent-infant attachment." *Psychiatry* (Edgmont) 5(8): 28–36.

Swain, J. E., J. P. Lorberbaum, et al. (2007). "Brain basis of early parent-infant interactions: Psychology, physiology, and in vivo functional neuroimaging studies." *J Child Psychol Psychiatry* 48(3–4): 262–87.

Swann, J. M., J. Wang, et al. (2003). "The MPN mag: Introducing a critical area mediating pheromonal and hormonal regulation of male sexual behavior." *Ann N Y Acad Sci* 1007: 199–210.

Symonds, T., M. Perelman, et al. (2007). "Further evidence of the reliability and validity of the premature ejaculation diagnostic tool." *Int J Impot Res* 19(5): 521–25.

Szinovacz, M. E. (1998a). "Grandparents today: A demographic profile." *Gerontologist* 38(1): 37–52.

Szinovacz, M. E., ed. (1998b). *Handbook on Grandparenthood.* Westport, CT: Greenwood.

Tamir, M., C. Mitchell, et al. (2008). "Hedonic and instrumental motives in anger regulation." *Psychol Sci* 19(4): 324–28.

Tamis-LeMonda, C. S., N. Cabrera, et al. (2002). *Handbook of Father Involvement: Multidisciplinary Perspectives.* Newark, NJ: Lawrence Erlbaum.

Tanagho, E. (2000). *Smith's General Urology.* London: McGraw-Hill.

Tannen, D. (1995). "[Varying styles of communication in men and women: . . . then you better bite your tongue]." *Krankenpfl Soins Infirm* 88(5): 1–3.

Tannen, D. (1990a). "Gender differences in topical coherence: Creating involvement in best friends' talk." *Discourse Processes,* special issue, *Gender and conversational interaction* 13(1): 73–90.

Tannen, D., ed. (2003). *Talking Past One Another: "But What Do You Mean?"—Women and Men in Conversation.* New York: Free Press.

Tannen, D., ed. (2001). *But What Do You Mean? Women and Men in Conversation.* New York: Free Press.

Tannen, D., ed. (1999). *The Power of Talk: Who Gets Heard and Why.* Boston: Irwin/The McGraw-Hill Companies.

Tannen, D., ed. (1993a). *Gender and Conversational Interaction.* New York: Oxford University Press.

Tannen, D., ed. (1993b). *The Relativity of Linguistic Strategies: Rethinking Power and Solidarity in Gender and Dominance.* New York: Oxford University Press.

Tannen, D., ed. (1990b). *Gender Differences in Conversational Coherence: Physical Alignment and Topical Cohesion.* Westport, CT: Ablex.

Tannen, D., and E. Aries, eds. (1997). *Conversational Style: Do Women and Men Speak Different Languages?* New Haven: Yale University Press.

Taylor, S. E., L. C. Klein, et al. (2000). "Biobehavioral responses to stress in females: Tend-and-befriend, not fight-or-flight." *Psychol Rev* 107(3): 411–29.

Teicher, M. D. (2000). "Wounds that time won't heal: The neurobiology of child abuse." *Cerebrum: The Dana Forum on Brain Science* 2(4): 50–67.

Teixeira, C., B. Figueiredo, et al. (2010). "Anxiety and depression during pregnancy in women and men." *J Affect Disord* 119(1–3): 142–48.

Teixeira, J., S. Maheswaran, et al. (2001). "Müllerian inhibiting substance: An instructive developmental hormone with diagnostic and possible therapeutic applications." *Endocr Rev* 22(5): 657–74.

Terburg, D., J. S. Peper, et al. (2009). "Sex differences in human aggression: The interaction between early developmental and later activational testosterone." *Behav Brain Sci* 32(3–4): 290; discussion 292–311.

Terlecki, M. S., and N. S. Newcombe (2005). "How important is the digital divide? The relation of computer and videogame usage to gender differences in mental rotation ability." *Sex Roles* 53(5–6): 433–41.

Terlecki, M. S., N. S. Newcombe, et al. (2008). "Durable and generalized effects of spatial experience on mental rotation: Gender differences in growth patterns." *Applied Cognitive Psychology* 22(7): 996–1013.

Tessitore, A., A. R. Hariri, et al. (2005). "Functional changes in the activity of brain regions underlying emotion processing in the elderly." *Psychiatry Res* 139(1): 9–18.

Thakkar, K. N., P. Brugger, et al. (2009). "Exploring empathic space: Correlates of perspective transformation ability and biases in spatial attention." *PLoS One* 4(6): e5864.

Thioux, M., V. Gazzola, et al. (2008). "Action understanding: How, what and why." *Curr Biol* 18(10): R431–R434.

Thomas, L. E., and A. Lleras (2009). "Swinging into thought: Directed movement guides insight in problem solving." *Psychon Bull Rev* 16(4): 719–23.

Thompson, M. E., and R. W. Wrangham (2008). "Male mating interest varies with female fecundity in *Pan troglodytes schweinfurthii* of Kanyawara, Kibale National Park." *International Journal of Primatology* 29(4): 885–905.

Thompson, R., S. Gupta, et al. (2004). "The effects of vasopressin on human facial responses related to social communication." *Psychoneuroendocrinology* 29(1): 35–48.

Thompson, R. R., K. George, et al. (2006). "Sex-specific influences of vasopressin on human social communication." *Proc Natl Acad Sci USA* 103(20): 7889–94.

Thomson, R. (2006). "The effect of topic of discussion on gendered language in computer-mediated communication discussion." *Journal of Language and Social Psychology* 25(2): 167–78.

Thornhill, R., and S. W. Gangestad (2008). *The Evolutionary Biology of Human Female Sexuality*. New York: Oxford University Press.

Thornhill, R., and S. W. Gangestad (1999). "The scent of symmetry: A human sex pheromone that signals fitness?" *Evol Hum Behav* 20: 175–201.

Tiemeier, H., R. K. Lenroot, et al. (2010). "Cerebellum development during childhood and adolescence: A longitudinal morphometric MRI study." *Neuroimage* 49(1): 63–70.

Timonin, M. E., and K. E. Wynne-Edwards (2008). "Aromatase inhibition during adolescence reduces adult sexual and paternal behavior in the biparental dwarf hamster *Phodopus campbelli*." *Horm Behav* 54(5): 748–57.

Tomaszycki, M. L., J. E. Davis, et al. (2001). "Sex differences in infant rhesus macaque separation-rejection vocalizations and effects of prenatal androgens." *Horm Behav* 39(4): 267–76.

Tomaszycki, M. L., H. Gouzoules, et al. (2005). "Sex differences in juvenile rhesus macaque (*Macaca mulatta*) agonistic screams: Life history differences and effects of prenatal androgens." *Dev Psychobiol* 47(4): 318–27.

Tommasi, L., and L. Nadel (2009). *Cognitive Biology: Evolutionary and Developmental Perspectives on Mind, Brain, and Behavior*, Vienna Series in Theoretical Biology. Vienna: Springer.

Tower, J. (2006). "Sex-specific regulation of aging and apoptosis." *Mech Ageing Dev* 127(9): 705–18.

Townsend, J. M., and T. Wasserman (1997). "The perception of sexual attractiveness: Sex differences in variability." *Arch Sex Behav* 26(3): 243–68.

Trainor, B. C., I. M. Bird, et al. (2004). "Opposing hormonal mechanisms of aggression revealed through short-lived testosterone manipulations and multiple winning experiences." *Horm Behav* 45(2): 115–21.

Trainor, B. C., I. M. Bird, et al. (2003). "Variation in aromatase activity in the medial preoptic area and plasma progesterone is associated with the onset of paternal behavior." *Neuroendocrinology* 78(1): 36–44.

Trainor, B. C., H. H. Kyomen, et al. (2006). "Estrogenic encounters: How interactions between aromatase and the environment modulate aggression." *Front Neuroendocrinol* 27(2): 170–79.

Trainor, B. C., and C. A. Marler (2002). "Testosterone promotes paternal behaviour in a monogamous mammal via conversion to oestrogen." *Proc Biol Sci* 269(1493): 823–29.

Trainor, B. C., and C. A. Marler (2001). "Testosterone, paternal behavior, and aggression in the monogamous California mouse (*Peromyscus californicus*)." *Horm Behav* 40(1): 32–42.

Trivers, R., J. Manning, et al. (2006). "A longitudinal study of digit ratio (2D:4D) and other finger ratios in Jamaican children." *Horm Behav* 49(2): 150–56.

Truitt, W. A., and L. M. Coolen (2002). "Identification of a potential ejaculation generator in the spinal cord." *Science* 297(5586): 1566–69.

Truitt, W. A., M. T. Shipley, et al. (2003). "Activation of a subset of lumbar spinothalamic neurons after copulatory behavior in male but not female rats." *J Neurosci* 23(1): 325–31.

Tsujimura, A., Y. Miyagawa, et al. (2009). "Sex differences in visual attention to sexually explicit videos: A preliminary study." *J Sex Med* 6(4): 1011–17.

Tsujimura, A., Y. Miyagawa, et al. (2006). "Brain processing of audiovisual sexual stimuli inducing penile erection: A positron emission tomography study." *J Urol* 176(2): 679–83.

Tsunematsu, T., L. Y. Fu, et al. (2008). "Vasopressin increases locomotion through a V1a receptor in orexin/hypocretin neurons: Implications for water homeostasis." *J Neurosci* 28(1): 228–38.

Tuljapurkar, S. (2009). "Demography: Babies make a comeback." *Nature* 460(7256): 693–94.

Tuljapurkar, S. D., C. O. Puleston, et al. (2007). "Why men matter: Mating patterns drive evolution of human lifespan." *PLoS One* 2(8): e785.

Tuman, D. M. (1999a). "Gender style as form and content in children's drawings." *Studies in Art Education* 41(1): 40–60.

Tuman, D. M. (1999b). "Sing a song of sixpence: An examination of sex differences in the subject preference of children's drawings." *Visual Arts Research* 25(1)[49]: 51–62.

Tyre, P. (2008). *The Trouble with Boys.* New York: Crown.

Tzur, G., and A. Berger (2009). "Fast and slow brain rhythms in rule/expectation violation tasks: Focusing on evaluation processes by excluding motor action." *Behav Brain Res* 198(2): 420–28.

Ullman, M. T. I., and J. B. Becker (2008). "Sex differences in the neurocognition of language." In: J. B. Becker, K. Berkley, N. Geary, E. Hampson, J. P. Herman, and E. A. Young, eds., *Sex Differences in the Brain: From Genes to Behavior.* Oxford, UK: Oxford University Press.

Unkelbach, C., A. J. Guastella, et al. (2008). "Oxytocin selectively facilitates recognition of positive sex and relationship words." *Psychol Sci* 19(11): 1092–94.

Updegraff, K. A., A. Booth, et al. (2006). "The role of family relationship quality and testosterone levels in adolescents' peer experiences: A biosocial analysis." *J Fam Psychol* 20(1): 21–29.

Vaglio, S., P. Minicozzi, et al. (2009). "Volatile signals during pregnancy: A possible chemical basis for mother-infant recognition." *J Chem Ecol* 35(1): 131–39.

Vaillancourt, T., D. deCatanzaro, et al. (2009). "Androgen dynamics in the context of children's peer relations: An examination of the links between testosterone and peer victimization." *Aggress Behav* 35(1): 103–13.

Vaillancourt, T., J. L. Miller, et al. (2007). "Trajectories and predictors of indirect aggression: Results from a nationally representative longitudinal study of Canadian children aged 2–10." *Aggress Behav* 33(4): 314–26.

Vaillant, G. E. (2002). *Aging Well.* Boston: Little, Brown.

Vale, J. R., D. Ray, et al. (1974). "Neonatal androgen treatment and sexual behavior in males of three inbred strains of mice." *Dev Psychobiol* 7(5): 483–88.

van Bokhoven, I., S. H. van Goozen, et al. (2006). "Salivary testosterone and aggression, delinquency, and social dominance in a population-based longitudinal study of adolescent males." *Horm Behav* 50(1): 118–25.

van Bokhoven, I., S. H. Van Goozen, et al. (2005). "Salivary cortisol and aggression in a population-based longitudinal study of adolescent males." *J Neural Transm* 112(8): 1083–96.

van der Meij, L., A. P. Buunk, et al. (2008). "The presence of a woman increases testosterone in aggressive dominant men." *Horm Behav* 54(5): 640–44.

van Eimeren, T., T. Wolbers, et al. (2006). "Implementation of visuospatial cues in response selection." *Neuroimage* 29(1): 286–94.

van Honk, J., J. S. Peper, et al. (2005). "Testosterone reduces unconscious fear but not consciously experienced anxiety: Implications for the disorders of fear and anxiety." *Biol Psychiatry* 58(3): 218–25.

van Honk, J., and D. J. Schutter (2007). "Testosterone reduces conscious detection of signals serving social correction: Implications for antisocial behavior." *Psychol Sci* 18(8): 663–67.

van Honk, J., D. J. Schutter, et al. (2004). "Testosterone shifts the balance between sensitivity for punishment and reward." *Psychoneuroendocrinology* 29(7): 937–43.

van Honk, J., A. Tuiten, et al. (2001). "A single administration of testosterone induces cardiac accelerative responses to angry faces in healthy young women." *Behav Neurosci* 115(1): 238–42.

van Nas, A., D. Guhathakurta, et al. (2009). "Elucidating the role of gonadal hormones in sexually dimorphic gene coexpression networks." *Endocrinology* 150(3): 1235–49.

Van Strien, J. W., R. F. Weber, et al. (2009). "Higher free testosterone level is associated with faster visual processing and more flanker interference in older men." *Psychoneuroendocrinology* 34(4): 546–54.

Veenema, A. H., and I. D. Neumann (2009). "Maternal separation enhances offensive play-fighting, basal corticosterone and hypothalamic vasopressin mRNA expression in juvenile male rats." *Psychoneuroendocrinology* 34(3): 463–67.

Veenema, A. H., and I. D. Neumann (2008). "Central vasopressin and oxytocin release: Regulation of complex social behaviours." *Prog Brain Res* 170: 261–76.

Vella, E. T., C. C. Evans, et al. (2005). "Ontogeny of the transition from killer to caregiver in dwarf hamsters (*Phodopus campbelli*) with biparental care." *Dev Psychobiol* 46(2): 75–85.

Vermeulen, A., S. Goemaere, et al. (1999). "Testosterone, body composition and aging." *J Endocrinol Invest* 22(5 suppl.): 110–16.

Vesterlund, L. (2008). "Gender differences in competition." *Negotiation Journal* 24: 447–64.

Vesterlund, L. (2007). "Do women shy away from competition? Do men compete too much?" *Quarterly Journal of Economics* 122(3): 1067–1101.

Viau, V. (2002). "Functional cross-talk between the hypothalamic-pituitary-gonadal and -adrenal axes." *J Neuroendocrinol* 14(6): 506–13.

Vincent, N. (2006). *Self-Made Man: One Woman's Journey into Manhood and Back Again.* New York: Viking.

Viviani, D., and R. Stoop (2008). "Opposite effects of oxytocin and vasopressin on the emotional expression of the fear response." *Prog Brain Res* 170: 207–18.

Voracek, M., and M. L. Fisher (2006). "Success is all in the measures: Androgenousness, curvaceousness, and starring frequencies in adult media actresses." *Arch Sex Behav* 35(3): 297–304.

Voyer, D., and J. Flight (2001). "Gender differences in laterality on a dichotic task: The influence of report strategies." *Cortex* 37(3): 345–62.

Waddell, J., D. A. Bangasser, et al. (2008). "The basolateral nucleus of the amygdala is necessary to induce the opposing effects of stressful experience on learning in males and females." *J Neurosci* 28(20): 5290–94.

Waldherr, M., and I. D. Neumann (2007). "Centrally released oxytocin mediates mating-induced anxiolysis in male rats." *Proc Natl Acad Sci U S A* 104(42): 16681–84.

Waldinger, M. D., P. Quinn, et al. (2005). "A multinational population survey of intravaginal ejaculation latency time." *J Sex Med* 2(4): 492–97.

Wallen, K. (2005). "Hormonal influences on sexually differentiated behavior in nonhuman primates." *Front Neuroendocrinol* 26(1): 7–26.

Wallen, K., and J. M. Hassett (2009). "Sexual differentiation of behaviour in monkeys: Role of prenatal hormones." *J Neuroendocrinol* 21(4): 421–26.

Walter, C. (2008). "Affairs of the lips: Why we kiss." *Scientific American, Mind* (February).

Walum, H., L. Westberg, et al. (2008). "Genetic variation in the vasopressin receptor 1a gene (AVPR1A) associates with pair-bonding behavior in humans." *Proc Natl Acad Sci U S A* 105(37): 14153–56.

Wang, C., E. Nieschlag, et al. (2009a). "Investigation, treatment and monitoring of late-onset hypogonadism in males." *Int J Androl* 32(1): 1–10.

Wang, C., E. Nieschlag, et al. (2009b). "ISA, ISSAM, EAU, EAA and ASA recommendations: Investigation, treatment and monitoring of late-onset hypogonadism in males." *Int J Impot Res* 21(1): 1–8.

Wang, P. Y., K. Koishi, et al. (2005). "Müllerian inhibiting substance acts as a motor neuron survival factor in vitro." *Proc Natl Acad Sci U S A* 102(45): 16421–25.

Wang, P. Y., A. Protheroe, et al. (2009). "Müllerian inhibiting substance contributes to sex-linked biases in the brain and behavior." *Proc Natl Acad Sci U S A* 106(17): 7203–8.

Wang, Z., and B. J. Aragona (2004). "Neurochemical regulation of pair bonding in male prairie voles." *Physiol Behav* 83(2): 319–28.

Wang, Z., and G. J. De Vries (1993). "Testosterone effects on paternal behavior and vasopressin immunoreactive projections in prairie voles (*Microtus ochrogaster*)." *Brain Res* 631(1): 156–60.

Warren, M. F., M. J. Serby, et al. (2008). "The effects of testosterone on cognition in elderly men: A review." *CNS Spectr* 13(10): 887–97.

Wasserman, G. A., and M. Lewis (1985). "Infant sex differences: Ecological effects." *Sex Roles* 12(5–6): 665–75.

Weber, B. J., and S. A. Huettel (2008). "The neural substrates of probabilistic and intertemporal decision making." *Brain Res* 1234: 104–15.

Wedekind, C., T. Seebeck, et al. (1995). "MHC-dependent mate preferences in humans." *Proc Biol Sci* 260(1359): 245–49.

Weinberg, M. K. T., Z. Edward, J. F. Cohn, and K. L. Olson (1999). "Gender differences in emotional expressivity and self-regulation during early infancy." *Developmental Psychology* 35(1): 175–88.

Weisfeld, G. E. (1999). *Evolutionary Principles of Human Adolescence*. New York: Basic Books.

Weisfeld, G. E., T. Czilli, et al. (2003). "Possible olfaction-based mechanisms in human kin recognition and inbreeding avoidance." *J Exp Child Psychol* 85(3): 279–95.

Weisfeld, G. E., D. M. Muczenski, et al. (1987). "Stability of boys' social success among peers over an eleven-year period." *Contributions to Human Development* 18: 58–80.

Weiss, P., and S. Brody (2009). "Women's partnered orgasm consistency is associated with greater duration of penile-vaginal intercourse but not of foreplay." *J Sex Med* 6(1): 135–41.

Welling, L. L., B. C. Jones, et al. (2008). "Men report stronger attraction to femininity in women's faces when their testosterone levels are high." *Horm Behav* 54(5): 703–8.

Welling, L. L., B. C. Jones, et al. (2007). "Raised salivary testosterone in women is associated with increased attraction to masculine faces." *Horm Behav* 52(2): 156–61.

Wessells, H., T. F. Lue, et al. (1996). "Penile length in the flaccid and erect states: Guidelines for penile augmentation." *J Urol* 156(3): 995–97.

Weyers, P., A. Muhlberger, et al. (2009). "Modulation of facial reactions to avatar emotional faces by nonconscious competition priming." *Psychophysiology* 46(2): 328–35.

Wild, B., M. Erb, et al. (2001). "Are emotions contagious? Evoked emotions while viewing emotionally expressive faces: Quality, quantity, time course and gender differences." *Psychiatry Res* 102(2): 109–24.

Willcox, B. J., Q. He, et al. (2006). "Midlife risk factors and healthy survival in men." *JAMA* 296(19): 2343–50.

Williams, J. G., C. Allison, et al. (2008). "The Childhood Autism Spectrum Test (CAST): Sex differences." *Journal of Autism and Developmental Disorders* 38(9): 1731–39.

Williams, L. M., M. J. Barton, et al. (2005). "Distinct amygdala-autonomic arousal profiles in response to fear signals in healthy males and females." *Neuroimage* 28(3): 618–26.

Williams, M. A., and J. B. Mattingley (2006). "Do angry men get noticed?" *Curr Biol* 16(11): R402–R404.

Williamson, M., and V. Viau (2008). "Selective contributions of the medial preoptic nucleus to testosterone-dependent regulation of the paraventricular nucleus of the hypothalamus and the HPA axis." *Am J Physiol Regul Integr Comp Physiol* 295(4): R1020–R1030.

Williamson, M., and V. Viau (2007). "Androgen receptor expressing neurons that project to the paraventricular nucleus of the hypothalamus in the male rat." *J Comp Neurol* 503(6): 717–40.

Winking, J., M. Gurven, et al. (2010). "The goals of direct paternal care among a South Amerindian population." *Am J Phys Anthropol* 139(3): 295–304.

Winking, J., H. Kaplan, et al. (2007). "Why do men marry and why do they stray?" *Proc Biol Sci* 274(1618): 1643–49.

Winslow, J. T., N. Hastings, et al. (1993). "A role for central vasopressin in pair bonding in monogamous prairie voles." *Nature* 365(6446): 545–48.

Wirth, M. M., and O. C. Schultheiss (2007). "Basal testosterone moderates responses to anger faces in humans." *Physiol Behav* 90(2–3): 496–505.

Wiszewska, A., Pawlowski, B. Boothroyd (2007). "Father-daughter relationship as a moderator of sexual imprinting: A facialmetric study." *Evolution and Human Behavior* 28(4): 248–52.

Witelson, S. F. (1991a). "Neural sexual mosaicism: Sexual differentiation of the human temporo-parietal region for functional asymmetry." *Psychoneuroendocrinology* 16(1–3): 131–53.

Witelson, S. F. (1991b). "Sex differences in neuroanatomical changes with aging." *N Engl J Med* 325(3): 211–12.

Witelson, S. F. (1989). "Hand and sex differences in the isthmus and genu of the human corpus callosum: A postmortem morphological study." *Brain* 112 (pt. 3): 799–835.

Witelson, S. F., H. Beresh, et al. (2006). "Intelligence and brain size in 100 postmortem brains: Sex, lateralization and age factors." *Brain* 129(pt. 2): 386–98.

Witelson, S. F., D. L. Kigar, et al. (2008). "Corpus callosum anatomy in right-handed homosexual and heterosexual men." *Arch Sex Behav* 37(6): 857–63.

Wolbers, T., E. D. Schoell, et al. (2006). "The predictive value of white matter organization in posterior parietal cortex for spatial visualization ability." *Neuroimage* 32(3): 1450–55.

Wood, G. E., and T. J. Shors (1998). "Stress facilitates classical conditioning in males, but impairs classical conditioning in females through activational effects of ovarian hormones." *Proc Natl Acad Sci U S A* 95(7): 4066–71.

Wood, J. L., V. Murko, et al. (2008). "Ventral frontal cortex in children: Morphology, social cognition and femininity/masculinity." *Soc Cogn Affect Neurosci* 3(2): 168–76.

Worthman, C. M. (2010). "Habits of the heart: Life history and the developmental neuroendocrinology of emotion." *Am J Hum Biol* 21(6): 772–81.

Wrangham, R., ed. (2006a). *Why Apes and Humans Kill.* New York: Cambridge University Press.

Wrangham, R. W., and M. L. Wilson, eds. (2004). *Collective Violence: Comparisons Between Youths and Chimpanzees.* New York: New York Academy of Sciences.

Wrangham, R. W., M. L. Wilson, et al. (2006b). "Comparative rates of violence in chimpanzees and humans." *Primates* 47(1): 14–26.

Wright, C. L., S. R. Burks, et al. (2008). "Identification of prostaglandin E2 receptors mediating perinatal masculinization of adult sex behavior and neuroanatomical correlates." *Dev Neurobiol* 68(12): 1406–19.

Wu, M. V., D. S. Manoli, E. J. Fraser, J. K. Coats, J. Tollkuhn, S.-I. Honda, N. Harada, and N. M. Shah (2010). "Estrogen masculinizes neural pathways and sex-specific behaviors." *Cell* 139(1): 61–72.

Wudy, S. A., H. G. Dorr, et al. (1999). "Profiling steroid hormones in amniotic fluid of midpregnancy by routine stable isotope dilution/gas chromatography-mass spectrometry: Reference values and concentrations in fetuses at risk for 21-hydroxylase deficiency." *J Clin Endocrinol Metab* 84(8): 2724–28.

Wyart, C., W. W. Webster, et al. (2007). "Smelling a single component of male sweat alters levels of cortisol in women." *J Neurosci* 27(6): 1261–65.

Wylie, K. R., and I. Eardley (2007). "Penile size and the 'small penis syndrome.' " *BJU Int* 99(6): 1449–55.

Wynne-Edwards, K. E. (2001). "Hormonal changes in mammalian fathers." *Horm Behav* 40(2): 139–45.

Wynne-Edwards, K. E., and C. J. Reburn (2000). "Behavioral endocrinology of mammalian fatherhood." *Trends Ecol Evol* 15(11): 464–68.

Xue, G., Z. Lu, et al. (2009). "Functional dissociations of risk and reward processing in the medial prefrontal cortex." *Cereb Cortex* 19(5): 1019–27.

Yamagiwa, J. (2001). "Factors influencing the formation of ground nests by eastern lowland gorillas in Kahuzi-Biega National Park: Some evolutionary implications of nesting behavior." *J Hum Evol* 40(2): 99–109.

Yamamoto, Y., B. S. Cushing, et al. (2004). "Neonatal manipulations of oxytocin alter expression of oxytocin and vasopressin immunoreactive cells in the paraventricular nucleus of the hypothalamus in a gender-specific manner." *Neuroscience* 125(4): 947–55.

Yamazaki, K., and G. K. Beauchamp (2007). "Genetic basis for MHC-dependent mate choice." *Adv Genet* 59: 129–45.

Yang, C. F., C. K. Hooven, et al. (2007). "Testosterone levels and mental rotation performance in Chinese men." *Horm Behav* 51(3): 373–78.

Yang, C. Y., J. Decety, et al. (2009). "Gender differences in the mu rhythm during empathy for pain: An electroencephalographic study." *Brain Res* 1251: 176–84.

Yaniv, I., S. Choshen-Hillel, et al. (2009). "Spurious consensus and opinion revision: Why might people be more confident in their less accurate judgments?" *J Exp Psychol Learn Mem Cogn* 35(2): 558–63.

Yassin, A. A., F. Saad, et al. (2008). "Metabolic syndrome, testosterone deficiency and erectile dysfunction never come alone." *Andrologia* 40(4): 259–64.

Yeh, K. Y., H. F. Pu, et al. (2010). "Different subregions of the medial preoptic area are separately involved in the regulation of copulation and sexual incentive motivation in male rats: A behavioral and morphological study." *Behav Brain Res* 205(1): 219–25.

Yoshimoto, D., A. Shapiro, et al., eds. (2005). *Nonverbal Communication Coding Systems of Committed Couples*. New York: Oxford University Press.

Young, E. A., and J. B. Becker (2009a). "Perspective: Sex matters—gonadal steroids and the brain." *Neuropsychopharmacology* 34(3): 537–38.

Young, L. J. (2009b). "Being human: Love—neuroscience reveals all." *Nature* 457(7226): 148.

Young, L. J. (2008). "Sex differences in affiliative behavior and social bonding." In: J. B. Becker, K. Berkley, N. Geary, E. Hampson, J. P. Herman, and E. A. Young, eds., *Sex Differences in the Brain: From Genes to Behavior*. Oxford, UK: Oxford University Press.

Yu, Q., Y. Tang, et al. (2009). "Sex differences of event-related potential effects during three-dimensional mental rotation." *Neuroreport* 20(1): 43–47.

Yuan, J., Y. Luo, et al. (2009). "Neural correlates of the females' susceptibility to negative emotions: An insight into gender-related prevalence of affective disturbances." *Hum Brain Mapp* 30(11): 3676–86.

Yurgelun-Todd, D. (2007). "Emotional and cognitive changes during adolescence." *Curr Opin Neurobiol* 17(2): 251–57.

Zahn-Waxler, C., M. Radke-Yarrow, E. Wagner, and M. Chapman (1992). "Development of concern for others." *Developmental Psychology* 28: 126–36.

Zak, P. J., and J. A. Barraza (2009). "Empathy and collective action." Available at SSRN: http://ssrn.com/abstract=1375059.

Zak, P. J., R. Kurzban, et al. (2005). "Oxytocin is associated with human trustworthiness." *Horm Behav* 48(5): 522–27.

Zak, P. J., A. A. Stanton, et al. (2007). "Oxytocin increases generosity in humans." *PLoS One* 2(11): e1128.

Zaki, J., J. Weber, et al. (2009). "The neural bases of empathic accuracy." *Proc Natl Acad Sci U S A* 106(27): 11382–87.

Zaviacic, M., V. Sisovsky, et al. (2009). "Cosmetic perfumes vs. human pheromones (natural chemical scents) of the human female and male in signalling and performing context of their sexual behaviour." *Bratisl Lek Listy* 110(8): 472–75.

Zehr, J. L., B. J. Todd, et al. (2006). "Dendritic pruning of the medial amygdala during pubertal development of the male Syrian hamster." *J Neurobiol* 66(6): 578–90.

Zelbergeld, B. (1999). *The New Male Sexuality*. New York: Bantam.

Zhang, Z., V. Klyachko, et al. (2007). "Blockade of phosphodiesterase type 5 enhances rat neurohypophysial excitability and electrically evoked oxytocin release." *J Physiol* 584(pt. 1): 137–47.

Zhou, L., J. D. Blaustein, et al. (1994). "Distribution of androgen receptor immunoreactivity in vasopressin- and oxytocin-immunoreactive neurons in the male rat brain." *Endocrinology* 134(6): 2622–27.

Zhou, W., and D. Chen (2008). "Encoding human sexual chemosensory cues in the orbitofrontal and fusiform cortices." *J Neurosci* 28(53): 14416–21.

Ziegler, T. E., S. Jacoris, et al. (2004). "Sexual communication between breeding male and female cotton-top tamarins (*Saguinus oedipus*), and its relationship to infant care." *Am J Primatol* 64(1): 57–69.

Ziegler, T. E., S. L. Prudom, et al. (2006). "Pregnancy weight gain: Marmoset and tamarin dads show it too." *Biol Lett* 2(2): 181–83.

Ziegler, T. E., and C. T. Snowdon (2000). "Preparental hormone levels and parenting experience in male cotton-top tamarins, *Saguinus oedipus*." *Horm Behav* 38(3): 159–67.

Ziegler, T. E., K. F. Washabaugh, et al. (2004). "Responsiveness of expectant male cotton-top tamarins, *Saguinus oedipus*, to mate's pregnancy." *Horm Behav* 45(2): 84–92.

Ziegler, T. E., F. H. Wegner, et al. (2000). "Prolactin levels during the periparturitional period in the biparental cotton-top tamarin (*Saguinus oedipus*): Interactions with gender, androgen levels, and parenting." *Horm Behav* 38(2): 111–22.

Ziegler, T. E., F. H. Wegner, et al. (1996). "Hormonal responses to parental and nonparental conditions in male cotton-top tamarins, *Saguinus oedipus*, a New World primate." *Horm Behav* 30(3): 287–97.

Zitzmann, M. (2006). "Testosterone and the brain." *Aging Male* 9(4): 195–99.

Zuloaga, D. G., D. A. Puts, et al. (2008). "The role of androgen receptors in the masculinization of brain and behavior: What we've learned from the testicular feminization mutation." *Horm Behav* 53(5): 613–26.

INDEX

ABOUT THE AUTHOR

Louann Brizendine, M.D., a neuropsychiatrist at the University of California–San Francisco, is the founder of the Women's and Teen Girls' Mood and Hormone Clinic. She was previously on faculty at the Harvard Medical School and is a graduate of Yale University School of Medicine and the University of California–Berkeley in neurobiology. She lives in the San Francisco Bay Area with her husband and son.